Quantum Communication

The Physical Layer of Future Optical Networks

New Era Electronics: A Lecture Notes Series

Series Editors: Vijay Raghunathan *(Purdue University, USA)*
Muhammad Ashraf Alam *(Purdue University, USA)*
Mark S Lundstrom *(Purdue University, USA)*

Published

Vol. 2   *Quantum Communication:*
The Physical Layer of Future Optical Networks
by Mahdi Hosseini

Vol. 1   *Transistors!*
by Mark S Lundstrom

New Era Electronics: a Lecture Notes Series · Volume 2

Quantum Communication

The Physical Layer of Future Optical Networks

Mahdi Hosseini

Purdue University, USA

NEW JERSEY · LONDON · SINGAPORE · BEIJING · SHANGHAI · HONG KONG · TAIPEI · CHENNAI · TOKYO

Published by

World Scientific Publishing Co. Pte. Ltd.

5 Toh Tuck Link, Singapore 596224

USA office: 27 Warren Street, Suite 401-402, Hackensack, NJ 07601

UK office: 57 Shelton Street, Covent Garden, London WC2H 9HE

Library of Congress Control Number: 2023939015

British Library Cataloguing-in-Publication Data
A catalogue record for this book is available from the British Library.

New Era Electronics: A Lecture Notes Series — Vol. 2
QUANTUM COMMUNICATION
The Physical Layer of Future Optical Networks

ISBN 978-981-127-905-8 (hardcover)
ISBN 978-981-127-908-9 (paperback)
ISBN 978-981-127-906-5 (ebook for institutions)
ISBN 978-981-127-907-2 (ebook for individuals)

For any available supplementary material, please visit
https://www.worldscientific.com/worldscibooks/10.1142/13490#t=suppl

Desk Editor: Joseph Ang

Typeset by Stallion Press
Email: enquiries@stallionpress.com

To Zagros and Yalda.

About the Author

Mahdi Hosseini is Faculty of Electrical and Computer Engineering at Purdue University and Northwestern University. He completed his Ph.D. and postdoctoral studies at the Australian National University (2012) and Massachusetts Institute of Technology (2016), respectively, where he studied quantum interactions of light with room-temperature and laser-cooled atomic gases. His research group investigates rare-earth photonics and room-temperature light-atom interfaces for quantum optical communication and sensing. He is a recipient of the 2022 National Science Foundation Career Award and the Director of the IQPARC (iqparc.com), a DoD-funded institute on quantum education.

Preface

The intrinsic quantum characteristics of laser light make photonics an attractive subject for studying many peculiar non-classical effects, such as superposition, entanglement, and teleportation. Recent theoretical and experimental advancements achieved in the field of quantum information suggest the possibility of a technological revolution based on employing such quantum mechanical phenomena in everyday applications. Quantum sensing, secure communication, and quantum computation are among the future technologies extensively investigated in various platforms. In secure communication, reliable manipulation and processing of quantum properties of light are crucial steps for developing photonic networks where the security of transmitted information is guaranteed by the laws of quantum physics. Currently, various systems are being investigated to find a scalable platform for information processing that is compatible with the current technology.

These notes introduce the basic quantum processes and devices used in quantum communication applications. The boundary between classical and quantum physics, the quantization of electromagnetic field and its consequences, quantum electromagnetic and atomic physics, and their applications in quantum communication are discussed. The device and experimental aspects of quantum key distribution, quantum teleportation, quantum memories, entanglement distribution, and quantum repeaters are also discussed.

Contents

Chapter 1

Introduction and Basic Concepts

1.1. Introduction

This course introduces basic concepts and elements of a quantum communication system. In this chapter, we start by reviewing some of the core and relevant concepts in quantum mechanics. We then, in the second chapter, introduce some quantum resources, such as photons and atoms, describe their interaction, and study their role in the generation and processing of quantum optical information. In the last chapter, we will discuss the physical layers of a quantum communication network.

1.1.1. *Why Quantum Networks?*

There are three main applications we should start designing and engineering quantum networks:

(1) secure communication,
(2) distributed quantum computing,
(3) distributed quantum sensing.

To understand the need for a quantum communication network, we first need to review classical communication, and study its limitations. The foundations of the Internet were developed in the 1960s when high-performance personal computers were not yet available. The innovative and collective thinking of engineers, physicists, mathematicians, and psychologists led to the design of hardware and software for connecting large-scale computers and later personal computers. The creation of the Internet was the start of a

1

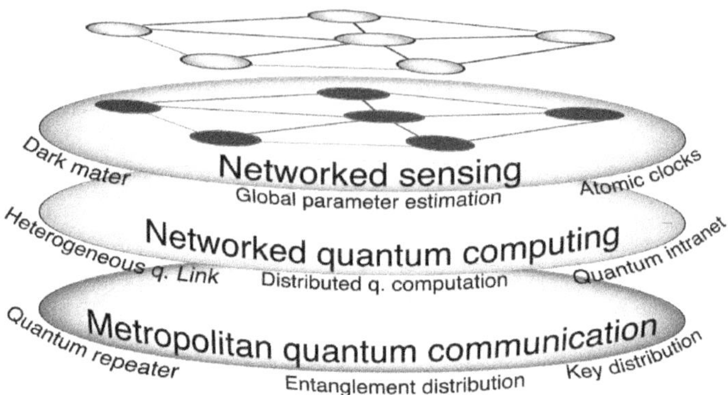

second digital revolution. The need for more secure communication protocols, the rise of quantum computing, the potential demand for connecting future quantum computers, and also the growing need for more precise sensors demand new communication hardware and software. The security of information and communication is becoming increasingly important in today's digital world. Novel communication protocols based on the laws of quantum mechanics can guarantee the security of information. For example, sharing quantum information encoded on entangled states of light can provide the means to securely communicate. Such communication networks could connect future quantum processors and sensors to perform computational and measurement tasks not possible with classical hardware.

1.1.2. *Light as a Precious Resource for Communication*

There are many reasons light is used as a reliable carrier of information, including:

- Fiber optics has revolutionized communication using light as a fast, noninterfering carrier.
- Photons can be digital and analogue at the same time (e.g. they can carry quantum information).
- Using quantum photons, secure communication is fundamentally guaranteed.

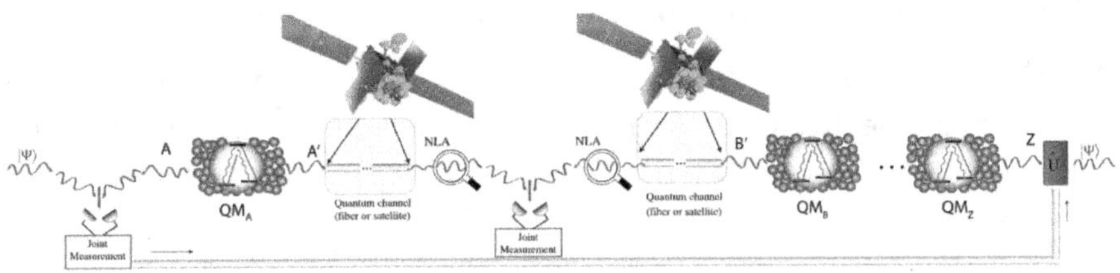

When it comes to communication, light is a very precious resource. It is evident how optical fibers and optical networks have revolutionized our communication in recent decades. The speed of light, lack of interaction and cross-coupling between photons, and ability of photons to carry information using their various degrees of freedom make light, or photons, an important carrier of information. Broadband communication can be achieved by sending information encoded into different degrees of freedom of light such as frequency. For these same reasons, we must consider photons as potential carriers of quantum information for future quantum networks. To reach the quantum regime and implement quantum communication, we shall define how light can be treated as a quantum mechanical object. A photon is a packet of optical energy in form of an electromagnetic field that travels at the speed of light. The energy within photons is quantized, and photons as quantum mechanical particles can exhibit both wave and particle natures. As we shall see, these important properties enable us to redefine communication by encoding information into a superposition or entangled states of light. Although information can be transferred through free space via satellites, for local and metropolitan networks, fiber-based networks are the preferred method of communication. In this document, we also discuss the role of quantum memories (QM) and joint measurements in quantum networks. An example of quantum communication links is shown in the above schematic diagram. Noiseless linear amplifiers (NLA) are also included in the above physical layer that may also be incorporated into quantum networks to purify entanglement. This is because quantum information degrades after transmission through quantum channels (fiber or satellite links) due to the inherent losses in the channels.

1.1.3. *Basic Concepts in Quantum Communication*

When "quantum encoding" is used, the laws of quantum physics guarantee the security of information. One form of quantum encoding is to use superposition of bits, instead of individual bits, to carry information

$$0 \ \& \ 1 \to |0\rangle + |1\rangle.$$

Quantum mechanics enables "teleportation" of information to distant locations. Assuming two distant nodes on the left (L) and right (R) ends of a network, a quantum state carrying information can be teleported from L to R node, without being directly transmitted between the two nodes:

$$A|0_L\rangle + B|1_L\rangle \longrightarrow A|0_R\rangle + B|1_R\rangle.$$

Distribution of entanglement is the basic need of a quantum network relying on teleportation for communication. Nodes of a network could form a linear chain along the communication channel or nodes could have multiple connections like the graph below.

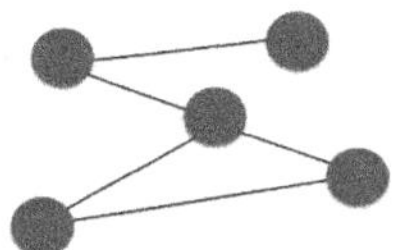

At the fundamental level, quantum communication relies on shared quantum information for connecting parties. A bit of information can be represented by numbers 0 or 1, however, in quantum mechanical language, a bit can be in a superposition of 0 and 1. We refer to this superposition of a 0 and a 1 as a "quantum bit" or a "qubit" to differentiate it from a classical bit. The superposition highlights the fact that there is a wave or a phase associated with qubits and they can interfere with one another. In other words, a qubit is both digital and analog. This is what fundamentally differentiates quantum systems from their classical counterparts.

As we shall see, quantum teleportation of information is possible. Teleportation requires the creation and distribution of entangled states, which are formed when two qubits are uniquely bonded together in a superposition state. We will see how interference followed by the detection of two quantum particles (e.g. an entangled photon, L, and a photon carrying information) can result in the mapping of quantum information to the other entangled qubit (photon B). This is called teleportation because quantum information carried by a qubit L is mapped to another qubit far away without directly sending the qubit L through the quantum channel. A teleportation is a powerful tool that has applications in distributed quantum computing. Entangled photons must be distributed to nodes of a network demanding the teleportation service. How these entangled photons are generated, distributed, stored, and measured are questions we try to answer in this document. We will focus on the optical and atomic-based hardware required to achieve such tasks.

1.2. Introduction to Quantum Mechanics

Firstly, we will introduce the main concepts and principles of quantum mechanics. We will use these principles and concepts to discuss the physics of quantum communication hardware. To begin mathematically describing a quantum mechanical system, we need to set some rules and principles. We refer to these principles as the postulates of quantum mechanics. These postulates allow us to define a mathematical framework for analyzing and describing quantum system dynamics.

1.2.1. *Postulates of Quantum Mechanics*

We first introduce some of the main concepts and principles of quantum mechanics most related to quantum communications. We only discuss a selected number of concepts and describe them in limited depth sufficient to understand the protocols and experiments described later in this chapter.

Postulate 1

A quantum system or particle is described by a wave function, $\psi(x, y, z, t)$, that contains information about the temporal and spatial properties of the system or the particle. The simplified wave function, $\psi(x, t)$, describes the system in a one-dimensional space and it does not consider change of or interaction with other degrees of freedom of the system.

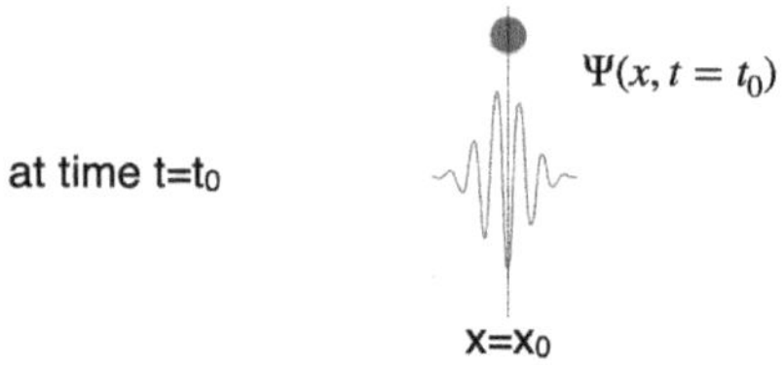

At its core, a quantum mechanical system is described by a "quantum state" or its "quantum wave function". A wave function is also called the probability amplitude of the system. Depending on the number of degrees of freedom of a particle that is considered, the wave function can have multiple dimensions including time. As a result, the wave function contains all the information that a particle or system can have. In practice, we may decide to consider or track a limited number of properties (or degrees of freedom) of a system most relevant to the problem at hand. The wave function of the system can have positive, negative, or imaginary values at certain locations in space and time. A wave function can evolve in time and space and can change due to interactions with other wave functions or processes. In many cases, we consider a simplified wave function where only one or a few degrees of freedom are tracked, assuming the other degrees of freedom are not affected, or they do not affect the outcome. Examples of such simplified states include the wave function of a particle oscillating or trapped in one dimension, the wave function of an electromagnetic field oscillating (polarized) along certain directions or the wave function of an electron or an atom with certain energies. A wave function has important properties, as we shall see when we discuss the remaining postulates, which describe both the particle and wave nature of a quantum system. These properties give rise to a complementary principle suggesting that although a wave function contains all information about the system, not all the properties of the system can be measured simultaneously.

Postulate 2

The probability density of a quantum mechanical object is given by the product of a wave function by its complex conjugate:

$\psi^*(x, y, z, t)\psi(x, y, z, t)$. To find the probability of funding a one-dimensional particle in region of space between x_1 and x_2, we calculate

$$\int_{x_1}^{x_2} \psi^*(x, t)\psi(x, t)dx$$

and total probability must add up to unity:

$$\int_{-\infty}^{\infty} \psi^*(x, t)\psi(x, t)dx = 1.$$

As stated, the wave function of a particle is the probability amplitude of the particle. To find the probability of detecting the particle in a region of space, we need to integrate the probability density over that space. The probability density for a one-dimensional (1D) particle has a unit of $[\mathrm{m}^{-1}]$ and is found by multiplying the wave function by its complex conjugate. A probability amplitude can be complex and negative, but the probability is measurable and has to be physical (real). Moreover, the total probability should be equal to one. This is the normalization condition of quantum wave functions, which states that the particle should exist somewhere in space and the total probability of events should add up to 1. A quantum wave function should always remain normalized, that is to say, its total probability of having certain properties should be one. This condition is used to find normalization coefficients of wave functions at different stages of the evolution or process. The probabilities associated with a system or particle can change upon interaction with the environment, however the total probability needs to be renormalized after each step of interaction and evolution. The concept of probability amplitude is unique to quantum mechanics and it enables us to explain fundamental differences between a classical system and a quantum one. In classical statistics, probabilities do not have amplitudes and they do not interfere. In quantum mechanics, probability amplitudes should be computed first and the probability afterward. As probability amplitudes are waves, they can destructively or constructively interfere. This may result in zero probability of finding particles in certain regions.

Postulate 3

For a particle in an isotropic medium, the wave function and its derivative should be continuous.

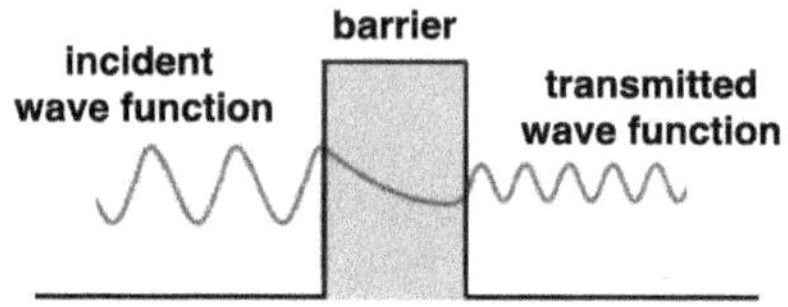

An important property of the wave function is that it is continuous. This property ensures that there is no singularity in space or time. We use this property to describe the wave function near its boundaries. This property also explains why a quantum mechanical object has a non-zero probability of going through a barrier whose potential is higher than the particle's energy. In quantum tunneling problems, where a quantum particle or a quantum wave encounters a barrier, the wave function's continuity indicates that the probability amplitude of finding the particle inside the barrier has to be non-zero to satisfy the continuity near the boundaries. This means that there is a non-zero probability that we can find the particle inside the barrier and consequently there is a non-zero probability that the particle will tunnel through the barrier. The lower the barrier's potential, the higher the tunneling probability. This postulate also indicates that the wave function derivative should be continuous. As we will see in the following sections, the derivative of the wave function can be related to the momentum or energy of the particle. Just like probability, these physical quantities cannot have singular values and must be continuous.

In some of the nano- and micro-photonic devices used to develop quantum communication systems, photons are confined in small regions of space using waveguides. Although photons do not leak outside the waveguide, it can be shown that the tail of the photon's wave function (i.e. evanescent field) can interact with atoms outside the waveguide.

Postulate 4

In quantum physics, "operators" take the place of dynamical variables in classical physics. The operators act on the wave function.

Some important examples of operators and the corresponding dynamical variables are shown in the following figure:

Dynamical variable in classical physics	Operator in quantum physics
x	x
f(x)	f(x)
p	$\dfrac{\hbar}{i}\dfrac{\partial}{\partial x}$
f(p)	$f\left(\dfrac{\hbar}{i}\dfrac{\partial}{\partial x}\right)$
E_{total}	$-\dfrac{\hbar}{i}\dfrac{\partial}{\partial t}$

To translate the problem from a classical picture to a quantum mechanical one, operators are used. We can associate an operator with each dynamical variable in classical physics. Operators act on the wave function and can alter it. Unlike classical variables, quantum mechanical operators are not physical quantities, but they enable us to calculate the average classical value of a measurement. Quantum mechanical operators are used to describe the effects of a physical process on the wave function. If the effect of applying an operator on a wave function is to modify the wave function with a scalar, then the wave function is called the eigenfunction and this scalar is called the eigenvalue. We express the eigenvalue equation as

$$\hat{O}\psi(x,r) = \lambda_o\psi(x,t).$$

The eigenvalue, λ_o, is a scalar constant.

In mathematical terms, the mathematical rule under which a wave function changes to another wave function or is simply scaled by a constant in the operation.

The wave function that an operator acts on is also called the operand. In the table above, $\hbar = h/2\pi$ and $h = 6.62607004 \times 10^{-34}\,\text{m}^2\,\text{kg/s}$ is the Planck constant. We shall see how these mathematical rules enable us to explain quantum mechanical phenomena, including the Heisenberg Uncertainty Principle.

Postulate 5

The expectation value of an operator, $\langle \hat{O} \rangle$, corresponds to the classical dynamical variable and it is calculated from the wave function according to

$$\langle \hat{O} \rangle = \int_{-\infty}^{\infty} \psi^*(x)\hat{O}\psi(x)dx.$$

To find the expected classical value corresponding to a quantum mechanical operator, we use the integral form presented above. The mean or expectation value of the operator is directly related to the physical quantity being observed. For instance, to find the expectation value of the energy of a particle using the quantum picture, we take the time derivative of the wave function (the operator equivalent to the classical variable energy as given in table above), multiply it by the complex conjugate of the wave function and then integrate the product state. The result of this integral is the mean energy of the particle that can be measured. To estimate the physical observable of interest such as position, velocity, energy, etc. of a particle, we need to calculate the expectation value and then we can compare it with the measured value (classical observable).

In optics, laser light can be described in the quantum picture by a wave function called the coherent state. The intensity of the laser is proportional to the number of photons in the laser beam. As we will see later, there is a corresponding operator for the observable: photon number. By computing the expectation value using the coherent state as a wave function and photon number operator, one can estimate the mean intensity or mean photon number of the laser beam. Note that every time the intensity of a laser is measured, it will show fluctuations around this mean value. The fluctuations are related to the probabilistic nature of the wave function, which we will formalize later in the context of the Heisenberg Uncertainty Principle.

1.2.2. *Hamiltonian and Schrödinger Equation*

In this section, we introduce some of the basic mathematical formalism used to describe quantum mechanical systems most relevant to quantum communication. We discuss how to write Hamiltonian of a quantum mechanical system and the corresponding Schrödinger equation to describe the evolution of the quantum wave function of the system.

We first provide a short summary about Schrödinger Equation and provide more details on the following paragraphs.

To describe the evolution of the wave function, Schrödinger introduced the following expression:

$$\hat{H}\psi(\boldsymbol{r},t) = i\hbar\frac{\partial}{\partial t}\psi(\boldsymbol{r},t).$$

The Hamiltonian, $\hat{H}$, is the quantum mechanical operator corresponding to the energy of the system and $\psi(\boldsymbol{r},t)$ is the generic three-dimensional wave function.

The Hamiltonian operator is of great importance because many problems of quantum mechanics are solved by minimizing the total energy of a particle or a system of particles.

In the early 20th century, Schrödinger introduced a differential equation to describe the dynamics of a quantum wave function. He defined the Hamiltonian operator corresponding to the dynamical variable energy, E. The wave function of a system with non-zero energy can undergo spatial and temporal variations. The above equation is the time-dependent Schrödinger equation that can be used to derive the equations of motion for different degrees of freedom captured by the wave function. This equation captures both the particle nature of a system as well as its wave nature. The equation assumes non-relativistic scenarios where the three spatial dimensions represented by **r** are independent of time. For simple systems involving few particles, the wave function and its evolution can be accurately described by this equation. For more complex systems involving several interacting particles, solving the Schrödinger equation can become extremely challenging. In fact, it is often not feasible.

We note that in the Schrödinger picture, it is the wave function that contains the time dependency of the system, so we write the time-dependent Schrödinger equation to extract the dynamics of the system. In contrast, the Heisenberg picture incorporates the time dependency into the operators. We will use both pictures in different examples to formulate quantum mechanical behavior.

Schrödinger

In one dimension, we can use position and momentum operator defined earlier to write the time-independent Hamiltonian of the system:

$$\hat{x} = x$$

$$\hat{p}_x = -i\hbar \frac{\partial}{\partial x}$$

$$\hat{H} = \frac{\hat{p}_x^2}{2m} + \hat{V}(x).$$

So, the 1D time-independent Schrödinger equation (eigenvalue equation) becomes

$$-\frac{\hbar^2}{2m} \frac{d^2 \psi(x)}{dx^2} + \hat{V}(x)\psi(x) = E\psi(x).$$

The Hamiltonian operator of a quantum mechanical system is analogous to the classical Hamiltonian written in terms of a Lagrangian:

$$H = \sum_j p_j \dot{q}_j - L$$

where q_j and p_j are the spatial coordinates and the canonical momentum (conjugate to q) of a particle or object, respectively. In the simplest form, for a 1D object, we can write $q_j = x$ and $p_j = p_x$ such that the classical Hamiltonian can be written as

$$H = \frac{p_x^2}{2m} + V(x)$$

where m is the mass of the object and $V(x)$ is the 1D-potential energy. This is nothing but the energy of the system. To obtain the

Hamiltonian operator, one can replace x and p_x with the corresponding quantum mechanical operators defined earlier. Therefore, the eigenvalue equation for the Hamiltonian operator, $\hat{H}\psi(x) = E\psi(x)$, can be written as shown above. Note that in this equation, the time evolution is not considered. This is valid if the potential energy is not time-independent. In this case, to describe the time evolution, one can use the time-dependent Schrödinger equation: $i\hbar\partial\psi(x,t)/\partial t = E\psi(x,t)$. Note that energy E here is the eigenvalue or expectation value of the Hamiltonian operator.

1.2.3. *Constant Potential*

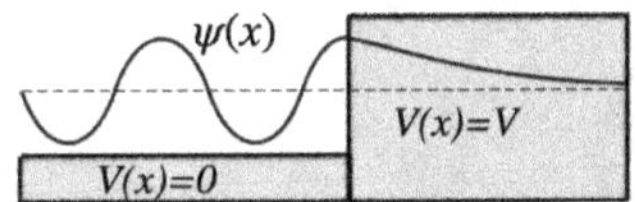

Assume a particle that moves in a potential $V(x) = $ constant. What are the possible solutions of the particle's wave function?

Using the Schrödinger equation above, we can write the following differential equation:

$$-\frac{\hbar^2}{2m}\frac{d^2\psi(x)}{dx^2} = (E - V)\psi(x).$$

The general solution to the above equation is written as

$$\psi(x) = Ae^{ikx} + Be^{-ikx}$$

where

$$k = \sqrt{\frac{2m}{\hbar^2}(E - V)}.$$

We consider a simple example where a particle or a wave with constant energy E bounces off a barrier modeled with a constant potential, V. Here, we only consider a one-dimensional space. We can write the eigenvalue equation for the particle's one-dimensional wave function, as shown above. This second-order differential equation has a known solution given by complex exponential functions, where the coefficient k can be real or imaginary. Depending on the energy of the particle compared to the potential energy $(E - V)$, the behavior of the particle can vary.

When $E > V$, the solution to the wave function is a sinusoidal oscillation in space. An oscillating wave that has real and imaginary components suggests it can propagate through the barrier. On the other hand, when $E < V$, k is imaginary resulting in a wave function with exponentially changing amplitude. The constants A and B can be found using the fact that the wave function must be continuous near the boundaries. We will provide more details about how to apply the continuity principle in another example later.

Practical examples of constant-potential problems in quantum mechanics include:

(1) quantum tunneling,
(2) quantum dot,
(3) reflection/transmission of EM field.

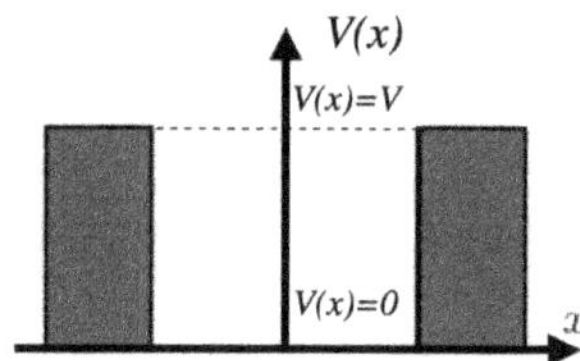

There are numerous examples of quantum mechanical systems where considering constant potential energies in different regions of space is sufficient to extract the main behavior of the system's wave function. For example, when considering a particle tunneling through a barrier with a constant potential, the solution for $\psi(x)$ described in the previous section captures the main features of quantum tunneling. The tunneling probability can then be calculated from the wave function. Scanning tunneling microscopes (STM), for example, rely on quantum tunneling of electrons from the surface to the imaging tip to detect the surface–tip distance. An electron trapped by a constant electric potential in a transistor (a quantum dot) can also be treated in this way to compute the probability of tunneling in and out of the transistor. Another example is an electromagnetic field of light reflected from a mirror, which can be treated quantum mechanically to find the field just outside the reflective mirror boundaries.

1.2.4. *Harmonic Oscillator*

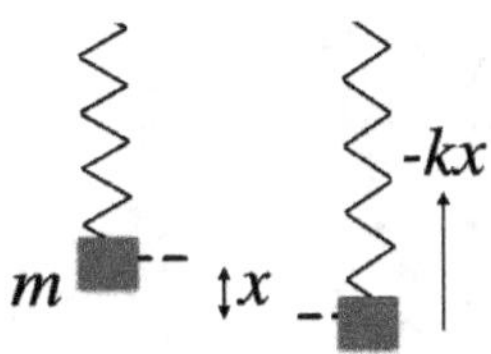

Let us now consider another example of a simple 1D quantum system that is ubiquitous in its application in solving quantum problems. A quantum particle bounded to a region of space via a restoring force is analogous to a classical particle of mass m attached to a spring with stiffness k. The potential energy is no longer constant and varies quadratically with position x. Nevertheless, the potential energy can be plugged into the time-independent Schrödinger equation to arrive at a new differential equation, as shown above.

For a particle or object oscillating in 1D on a spring, the potential energy is written as

$$V(x) = \frac{1}{2}kx^2 = \frac{1}{2}m\omega^2 x^2.$$

In the quantum regime, the corresponding 1D Schrödinger equation is

$$-\frac{\hbar^2}{2m}\frac{d^2\psi(x)}{dx^2} + \frac{1}{2}m\omega^2 x^2\psi(x) = E\psi(x).$$

The angular frequency $\omega = \sqrt{k/m}$. This differential equation has a known set of solutions. The solution is not unique and each solution corresponds to a specific energy, revealing the quantization of the particle's energy.

The solutions to the above differential equation are of the form

$$\psi_n(x) = u_n(x)\exp(-m\omega x^2/2\hbar)$$

where the Hermite polynomial functions, $u_n(x)$, are defined as

n	$u_n(x)$
0	$(1/a\sqrt{\pi})^{1/2}$
1	$(1/2a\sqrt{\pi})^{1/2}2(x/a)$
2	$(1/8a\sqrt{\pi})^{1/2}\left[2 - 4(x/a)^2\right]$
3	$(1/48a\sqrt{\pi})^{1/2}\left[12(x/a) - 8(x/a)^3\right]$
$\vdots$	
n	$(1/n!2^n a\sqrt{\pi})^{1/2} H_n(x/a)$

We see that the solution to the time-independent Schrödinger equation is given by a Gaussian function multiplied by a Hermit polynomial function shown above. Here, the length constant $a = \sqrt{\hbar/m\omega}$. For any integer value n, there exists a unique solution given by the polynomial function of order n. By substituting $\psi_n(x)$ back into the time-independent Schrödinger equation, we can find the possible energies of the particle that agree with these solutions.

Below we will see that the energy of a quantum harmonic oscillator possesses quantized values. The quantum number, n, characterizes the energy and the corresponding wave function of the system. For $n = 0$, the particle has the lowest energy allowed by quantum mechanics. This is called the zero-point or the ground-state energy of an oscillator. Later in this chapter, we will provide more details about this solution and its applications.

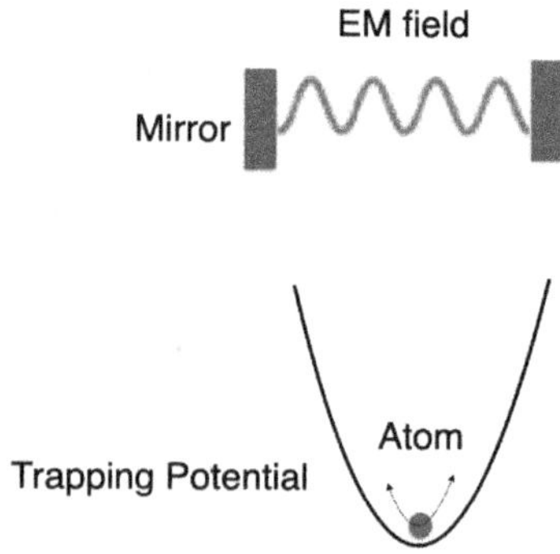

Like the constant-potential problem, the Harmonic oscillator problem appears over and over in quantum mechanics. Two examples

of that include the following: (1) a single-mode electromagnetic (EM) field oscillating between two mirrors or inside a cavity and (2) atoms trapped in a harmonic potential trap formed with laser beams.

In the first example, the Maxwell equations can be used to rewrite the energies of the electric and magnetic fields that resemble a harmonic oscillator. This analogy can be used to show the quantization of EM field energies.

Also, restoring laser forces can be used to create a harmonic trap for an atom localized and oscillating in space. The dynamics of the atom in the trap can be described in 1D or 3D using the harmonic potential model to obtain the ground and excited state oscillation energies as well as the wave function of the atom. At low thermal noise and close to the quantum ground state of the atom's motion, one can observe the wave effects of the atom's wave function in the trap (e.g. interference with other atoms). At very low energies near the quantum ground state of motion, the wave function is spatially extended. This picture can be used to understand the interaction between multiple atoms and the creation of Bose–Einstein condensation. The field of quantum optomechanics studies harmonic oscillation of objects (in quantum domain) as small as a single atom and as large as kilogram-large mirrors interfaced with light or photons.

1.2.5. *Quantization of a Harmonic Oscillator's Energy*

In the case of a quantum harmonic oscillator, we can plot $\psi_n(x)$, the solution to the 1D time-independent Schrödinger equation.

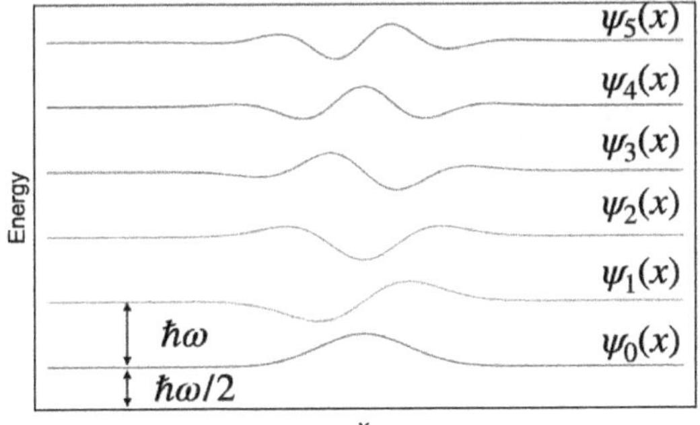

We saw that the solution to the Schrödinger equation is given by

$$\psi_n(x) = u_n(x)\exp(-m\omega x^2/2\hbar)$$

where

n	$u_n(x)$
0	$(1/a\sqrt{\pi})^{1/2}$
1	$(1/2a\sqrt{\pi})^{1/2}2(x/a)$
2	$(1/8a\sqrt{\pi})^{1/2}[2 - 4(x/a)^2]$
3	$(1/48a\sqrt{\pi})^{1/2}[12(x/a) - 8(x/a)^3]$
$\vdots$	
n	$(1/n!2^n a\sqrt{\pi})^{1/2}H_n(x/a)$

By plugging the solutions to the Schrödinger equation, we can show that energies of a harmonic oscillator are quantized:

$$E_n = \left(n + \frac{1}{2}\right)\hbar\omega.$$

The plot above shows the wave function for the first six energies. The simplest wave function belongs to the ground-state energy of the system with minimum energy $\hbar\omega/2$ (see next section). The difference between E_n and E_{n-1} is always $\hbar\omega$. However, when the oscillator approaches the classical regime, i.e. $n \gg 1$, the energy difference between the neighboring states becomes negligible compared to E_n. As a result, the quantization of energy can no longer be observed and classical theory is sufficient to describe the dynamics. In this regime, an oscillating particle does not show wave-like properties (e.g. interference) and an oscillating wave does not show particle properties (e.g. quantized energies). In the quantum regime, where $n \sim 1$, the wave nature of the particle can be observed. This means that the particle's wave function can be the result of interference (or superposition) between multiple eigenfunctions of the Hamiltonian operator. In other words, it is not just $\psi_n(x)$, but any linear combination of $\psi_n(x)$ which is a solution to the Schrödinger equation can be the solution to the system. The boundary between quantum and classical physics is not well defined. Identification of the boundary depends on many parameters, e.g. the precision of measurement instruments and the magnitude of loss and noise within the system. As our ability to measure and control mesoscopic and macroscopic systems increases, we can further push the boundary of the quantum–classical regime.

1.2.6. *Energies of a 1D Harmonic Oscillator*

In this section we calculate the ground-state energy of a 1D harmonic oscillator, oscillating along x direction with frequency ω and mass m. The total Hamiltonian is given by the sum of kinetic and potential energies substituting dynamical variables with operators

$$\hat{H} = \hat{T} + \hat{V} = \frac{\hat{P}_x^2}{2m} = -\frac{\hbar^2}{2m}\frac{\partial^2}{\partial x^2} + \frac{1}{2}m\omega^2\hat{x}^2$$

For a harmonic oscillator in the ground state energy, the wave function is given by

$$|\psi_0\rangle = \left(\frac{1}{a\sqrt{\pi}}\right)^{1/2} e^{-\frac{x^2}{2a^2}}$$

where we have defined $a = \sqrt{\hbar/m\omega}$. We can then write the time-independent Schrödinger equation

$$\hat{H}|\psi_0\rangle = \left(-\frac{\hbar^2}{2m}\frac{\partial^2}{\partial x^2} + \frac{1}{2}m\omega^2\hat{x}^2\right)|\psi_0\rangle = E|\psi_0\rangle$$

we then use the following operation

$$\frac{\partial^2}{\partial x^2}|\psi_0\rangle = \frac{\partial^2}{\partial x^2}\left\{\left(\frac{1}{a\sqrt{\pi}}\right)^{1/2} e^{-\frac{x^2}{2a^2}}\right\}$$

$$= \left(\frac{1}{a\sqrt{\pi}}\right)^{1/2}\left(-\frac{1}{a^2} + \frac{x^2}{a^4}\right) e^{-\frac{x^2}{2a^2}}$$

to rewrite the Schrödinger equation

$$\hat{H}|\psi_0\rangle = -\frac{\hbar^2}{2m}\left\{\left(\frac{1}{a\sqrt{\pi}}\right)^{1/2}\left(-\frac{1}{a^2} + \frac{x^2}{a^4}\right) e^{-\frac{x^2}{2a^2}}\right\}$$

$$+ \frac{1}{2}m\omega^2\hat{x}^2\left(\frac{1}{a\sqrt{\pi}}\right)^{1/2} e^{-\frac{x^2}{2a^2}}$$

$$= E\left(\frac{1}{a\sqrt{\pi}}\right)^{1/2} e^{-\frac{x^2}{2a^2}}$$

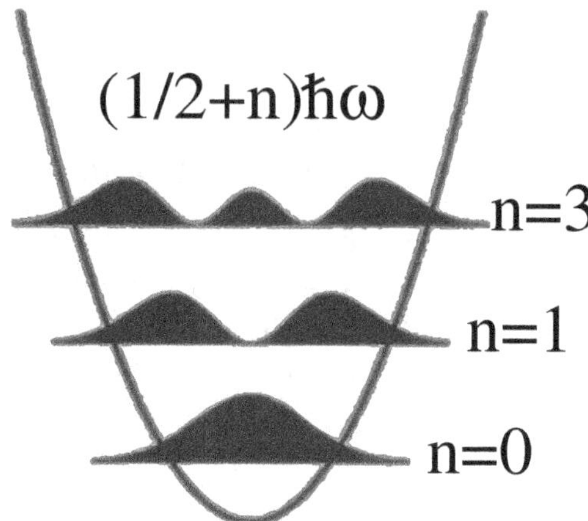

Illustration of the first three wave functions of a 1D Harmonic oscillator.

Canceling terms from both sides and simplifying further we have

$$E = -\frac{\hbar^2}{2m}\left(-\frac{1}{a^2}\right) = \frac{1}{2}\hbar\omega$$

This is the ground state energy of a 1D harmonic oscillator as expected. One can repeat the calculation for different $|\psi_n\rangle$ and find that

$$\langle \hat{H} \rangle = E_n = \left(\frac{1}{2} + n\right)\hbar\omega.$$

See the figure above.

1.2.7. *Example: Normalization*

Consider a particle in a harmonic potential with $V(x) = 1/2kx^2$ is initially in a superposition of ground state ($n = 0$) and first excited state ($n = 1$) energy.

$$\Psi(x, t = 0) = C[\psi_0(x, t = 0) + \psi_1(x, t = 0)]$$

We like to find C, assuming $\psi_0(x)$ and $\psi_1(x)$ are normalized.

The solutions to the harmonic oscillator equation are orthogonal wave functions given by

$$\psi_n(x) = u_n(x) \exp\left(-\frac{m\omega x^2}{2\hbar}\right)$$

$$\int_{-\infty}^{\infty} \psi_m^*(x)\psi_n(x)dx = \delta_{mn}$$

where the Dirac delta

$$\delta_{mn} = 0, \quad m \neq n$$

$$\delta_{mn} = 1, \quad m = n$$

The total wave function needs to be normalized:

$$\int_{-\infty}^{\infty} \Psi^*(x)\Psi(x)dx = 1 = C^2(1+1) \to C = \frac{1}{\sqrt{2}}.$$

We know that each $\psi_n(x)$ is orthonormal. However, when the solution is expressed as a superposition of eigenfunctions, it needs to be renormalized. In this case, we can find the normalization constant, C, by ensuring that the total probability adds up to 1. Taking into account that eigenfunctions with different n's do not overlap (they are orthogonal), we can calculate the integral to find C, as shown above. This normalization process needs to be repeated every time the particle undergoes a process that changes its resulting wave function.

An oscillating particle initially in its quantum ground state can be excited to the first energy state by absorbing one quanta of vibrational energy (a phonon) at its resonance frequency. However, if the phonon wave function is in a superposition of zero and one phonon energy, the particle can be transferred to a superposition state as well.

We shall see later that a low-noise laser beam is made of photons (quantized particles carrying electromagnetic energy) that are inherently in superposition states of different energies (i.e. superposition of $0, 1, 2, \ldots$ photons). The average intensity of a laser beam can be used to determine the probability of each photon-number state. At very low intensities, it is possible to virtually create a superposition of 0 and 1 photon only. As photons can be absorbed and their momentum can be transferred to oscillating particles, a photonic superposition can be used to create energy superposition in the particle.

1.2.8. *Example: Time Evolution*

A particle in a harmonic potential $V(x) = 1/2kx^2$ is initially in a superposition of ground state $(n = 0)$ and first excited state $(n = 1)$ energy:

$$\psi(x, t = 0) = C[\psi_0(x, t = 0) + \psi_1(x, t = 0)].$$

Here we like to find the time evolution of the wave function, $\psi(x, t)$.

To find the time-dependent wave function of the particle in the same example above, we need to use the time-dependent Schrödinger equation. This results in a first-order differential equation in time, which can be solved to find $\Psi(x, t)$ from $\Psi(x, t = 0)$.

We write the time-dependent 1D Schrödinger equation as

$$\hat{H}\psi_n(x, t) = i\hbar\frac{\partial}{\partial t}\psi_n(x, t)$$

Note that the eigenvalue equation is

$$\hat{H}\psi_n(x, t) = E_n\psi_n(x, t)$$

where

$$E_n = \left(n + \frac{1}{2}\right)\hbar\omega.$$

Thus,

$$E_n\psi_n(x, t) = i\hbar\frac{\partial}{\partial t}\psi_n(x, t) \longrightarrow \psi_n(x, t) = \psi_n(x)e^{-iE_n t/\hbar}$$

$$\Psi(x, t) = \frac{1}{\sqrt{2}}[\psi_0(x)e^{-iE_0 t/\hbar} + \psi_1(x)e^{-iE_1 t/\hbar}].$$

The final solution reveals an exponential form for the time-dependent part of the wave function. The term $E_n t/\hbar$ determines the phase of each eigenfunction. It can be seen that the two eigenfunctions constructing $\Psi(x, t)$ undergo interference that change in time. As a result, the total probability amplitude of the particle oscillates in time and can have negative values.

Interference of this kind reveals a fundamental difference between classical and quantum probabilities. In classical statistics, there is no need for considering probability amplitudes and therefore there is no interference between possible outcomes. Quantum probabilities, however, can interfere, leading to the enhancement or suppression of certain outcomes. This is only surprising when we consider both particle and wave natures of quantum mechanical systems. In classical wave theory, interference happens all the time. In a double-slit experiment with classical light, interference gives rise to dark fringes, suggesting a close-to-zero probability of light in certain locations in space. In the quantum picture, a photon or a single atom as a particle can interfere with itself just like a wave. Consider this question: how can a particle cross a one-dimensional space from point A to C without ever being seen at point B (located between A and C)?

1.2.9. *Example: Uncertainty in Energy of a Harmonic Oscillator*

A particle in a harmonic potential $V(x) = 1/2kx^2$ is initially in a superposition of ground state ($n = 0$) and first excited state ($n = 1$) energy.

$$\psi(x,t) = \frac{1}{\sqrt{2}}[\psi_0(x)e^{-iE_0t/\hbar} + \psi_1(x)e^{-iE_1t/\hbar}].$$

We like to calculate variation or standard deviation in its energy $\Delta E = \sqrt{\langle E^2 \rangle - \langle E \rangle^2}$. Using the derived wave function, now we would like to calculate the expectation values of E and E^2. These correspond to the average values of the energy and the square of energy. By calculating $\sqrt{\langle E^2 \rangle - \langle E \rangle^2}$, we can then find the standard deviation of the fluctuations in energy. In quantum mechanical terms, we refer to this as uncertainty in determining the energy of the system.

By plugging $\psi(x,t)$ in the expectation value equation, we can find the expressions for $\langle E^2 \rangle$ and $\langle E \rangle$ in terms of the energies of the systems, E_0 and E_1. From the definition of the expectation value of an operator, we can write

$$\langle E \rangle = \int_{-\infty}^{\infty} \psi^*(x,t) \hat{H} \Psi(x,t) dx$$

$$= \int_{-\infty}^{\infty} \psi^*(x,t) \left(i\hbar \frac{\partial}{\partial t} \right) \Psi(x,t) dx = \frac{1}{2}(E_0 + E_1).$$

Similarly,

$$\langle E^2 \rangle = \int_{-\infty}^{\infty} \psi^*(x,t) \left(i\hbar \frac{\partial}{\partial t} \right)^2 \Psi(x,t) dx = \frac{1}{2}(E_0^2 + E_1^2).$$

Thus,

$$\Delta E = \sqrt{\langle E^2 \rangle - \langle E \rangle^2} = \pm\frac{1}{2}(E_1 - E_0).$$

This is the uncertainty in measuring the energy of the state.

The uncertainty in measuring the system's energy is given in terms of $|1/2(E_1 - E_0)| = \hbar\omega/4$. We will introduce the notion of uncertainty again in the context of the Heisenberg Uncertainty Principle to see what minimum uncertainty a quantity can possess.

1.2.10. *Tunneling*

In the next example, we consider a potential barrier in a one-dimensional space and a particle with energy E traveling toward the barrier. The barrier is located between $0 < x < a$, where a is the width of the barrier. We are interested in finding out the probability of the particle tunneling through the barrier in terms of the system parameters.

Consider a particle with energy E traveling in 1D space facing rectangular potential barrier as shown in the following figure:

I: $V(x) = 0, \ x < 0$

II: $V(x) = U, \ 0 \leq x \leq a$

III: $V(x) = 0, \ x > a$

The total energy of the particle is given in terms of its kinetic energy (K.E.) and potential energy (P.E.) across different regions in space. We have already discussed that in this constant-potential problem, the generic solution to the time-dependent Schrödinger equation takes the form of an exponential function. We can find the constants A and B for each region of space using the continuity and normalization principles.

But first, let's write a solution for the wave function for each region of the space. In the regions I and III there is no potential energy and the total energy of the particle is simply the kinetic energy, E.

We can write the 1D Schrödinger equation as

$$\left[\frac{-\hbar^2}{2m}\frac{d^2}{dx^2} + V(x)\right]\psi(x) = E\psi(x)$$

$$\text{H} = \text{K.E.} + \text{P.E.}$$

For region I (and III), we have $V(x) = 0$:

$$\frac{d^2}{dx^2}\psi_I(x) = -\underbrace{\frac{2mE}{\hbar^2}}_{\mu^2}\psi_I(x) \xrightarrow[\text{solution}]{\text{general}} \psi_I(x) = A_I e^{i\mu x} + B_I e^{-i\mu x}$$

$$\text{sinusoidal oscillation}$$

In these regions, the exponents will be imaginary suggesting sinusoidal oscillations of the wave function in space. The spatial frequency (or wave number) in these regions is then given by $\mu = \pm\sqrt{2mE/\hbar^2}$. However, the coefficients A and B of the solutions in these two regions are not necessarily the same.

Let's now consider the solution in the region II, where the potential energy of the barrier is larger than the kinetic energy of the particle.

For region II, we have $V(x) = U$:

$$\frac{d^2}{dx^2}\psi_{\text{II}}(x) = \underbrace{\frac{2m(U-E)}{\hbar^2}}_{\lambda^2}\psi_{\text{II}}(x) \xrightarrow[\text{solution}]{\text{general}} \psi_{\text{II}}(x) = A_{\text{II}} e^{\lambda x} + B_{\text{II}} e^{-\lambda x}$$

$$\text{exponential decay}$$

The exponent, in this case, will be real with an exponential constant, $\lambda = \pm\sqrt{2m(U - E)/\hbar^2}$. It can be shown that the coefficient A_{II} must be zero to avoid a singularity and to satisfy the continuity conditions at the boundaries. Therefore, the solution represents an exponentially decaying wave function in the barrier region.

The non-zero probability amplitude suggests that there is a non-zero probability of finding a particle within the barrier region while it is tunneling through to the other side. This is counterintuitive and not consistent with the classical picture. Note that in limited cases, the two pictures agree and are consistent with each other. In the classical picture, a ball bouncing off a hard barrier (e.g. a wall) can never tunnel through the wall, but a quantum particle can do so with some probability. This means that if we repeat the experiment many times, occasionally the particle can be found on the other side of the barrier. But to find out the probability of such tunneling events and to see how often the particle may tunnel through the barrier, we need to find the coefficients A and B in terms of the physical parameters. Now that we have identified the form of the solutions to the wave function in different regions of space, we can find the coefficients of the solution using the quantum mechanic postulates: the wave function and its derivative have to be continuous in space and the total probability must be equal to unity, in other words, the wave function has to be normalized.

Summarizing the solutions above for the three regions, we have

$$\psi_{\mathrm{I}}(x) = A_{\mathrm{I}}e^{i\mu x} + B_{\mathrm{I}}e^{-i\mu x}$$

$$\psi_{\mathrm{II}}(x) = A_{\mathrm{II}}e^{\lambda x} + B_{\mathrm{II}}e^{-\lambda x}$$

$$\psi_{\mathrm{III}}(x) = A_{\mathrm{III}}e^{i\mu x} + B_{\mathrm{III}}e^{-i\mu x}.$$

To find the coefficients, we use the continuity and normalization conditions at the boundaries:

$$\psi_{\mathrm{I}}(x = 0) = \psi_{\mathrm{II}}(x = 0) \longrightarrow A_{\mathrm{I}} + B_{\mathrm{I}} = A_{\mathrm{II}} + B_{\mathrm{II}}$$

$$\frac{d}{dx}\psi_{\mathrm{I}}(x = 0) = \frac{d}{dx}\psi_{\mathrm{II}}(x = 0) \longrightarrow i\mu(A_{\mathrm{I}} - B_{\mathrm{I}}) = \lambda(A_{\mathrm{II}} - B_{\mathrm{II}})$$

$$\psi_{\mathrm{II}}(x = a) = \psi_{\mathrm{III}}(x = a) \longrightarrow A_{\mathrm{II}}e^{\lambda a} + B_{\mathrm{II}}e^{-\lambda a}$$

$$= A_{\mathrm{III}}e^{i\mu a} + B_{\mathrm{III}}e^{-i\mu a}$$

$$\frac{d}{dx}\psi_{\text{II}}(x=0) = \frac{d}{dx}\psi_{\text{III}}(x=0) \longrightarrow \lambda(A_{\text{II}}e^{\lambda a} - B_{\text{II}}e^{-\lambda a})$$

$$= i\mu(A_{\text{III}}e^{i\mu a} - B_{\text{III}}e^{-i\mu a}).$$

The four equations above are the result of applying the continuity condition to the wave function and its derivative at boundaries between regions I and II and II and III.

To ensure the wave function is normalized, we need to have the total probability equal to one as follows:

$$\int_{-\infty}^{\infty} \psi^*(x)\psi(x)dx = 1$$

$$= \int_{-\infty}^{0} \psi_{\text{I}}^*(x)\Psi_{\text{I}}(x)dx + \int_{0}^{a} \psi_{\text{II}}^*(x)\psi_{\text{II}}(x)dx + \int_{a}^{\infty} \psi_{\text{III}}^*(x)\psi_{\text{III}}(x)dx.$$

We also established that A_{II} has to be zero to avoid a singularity.

Overall, we can use the above-mentioned five equations to find the five remaining coefficients of the solution.

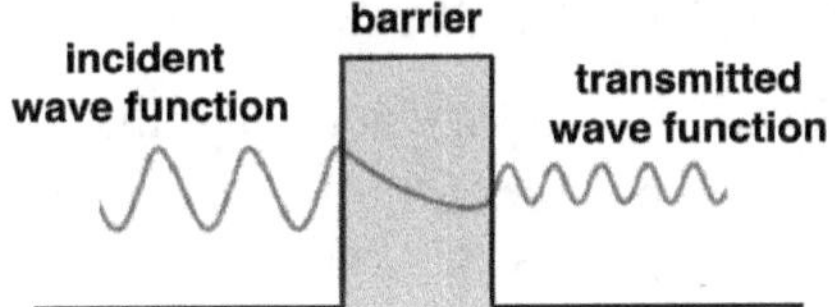

After solving for coefficients of the solution given above, we find the wave function (or probability amplitude) of the particle in different regions of the space. For example, A_{I} is the probability amplitude of finding a particle traveling from left to right while B_{I} is the probability amplitude of finding a particle traveling in the opposite direction in the region I. For practical purposes, the tunneling probability defined as the square of the ratio of $A_{\text{III}}/A_{\text{I}}$ is the relevant quantity of interest that can be measured experimentally.

The tunneling probability through the barrier is then given by

$$T = \left|\frac{A_{\text{III}}}{A_{\text{I}}}\right|^2 = \frac{4\lambda^2\mu^2}{(\mu^2 + \lambda^2)^2 \sinh^2(\lambda a) + 4\lambda^2\mu^2}.$$

The tunneling probability measures the likelihood of a particle being transmitted through the barrier. Therefore, the four equations derived from the continuity conditions are sufficient to calculate the four remaining coefficients relative to A_I. The tunneling probability, T, can be written as shown above. It depends on the kinetic energy of the particle, potential energy, and width of the barrier.

The tunneling problems appear in many branches of quantum physics and engineering. In explaining the radioactive decay process, for example, quantum tunneling enables us to estimate the escaping probability of α-particle from the nucleus. Superconducting circuits used in quantum computing rely on the quantum tunneling of electrons through an insulating barrier to create the nonlinearity required to realize and manipulate qubits. Field electron emission (emission of electrons induced by electromagnetic field) is another example, where quantum tunneling provides important insights.

In the field electron emission process, electrons escape the conducting surface and tunnel through an insulating barrier. As the tunneling probability is primarily defined by the insulating gap, the measured tunneling current can be used to estimate the barrier width.

Scanning tunneling microscope (STM)

The tunneling notion of a field electron emission gave rise to the invention of one of the most precise microscopes ever built, called the scanning tunneling microscope (STM). An STM works by measuring the tunneling current between a sharp metallic tip and the conductive surface being imaged. In this scenario, the tunneling probability can be further simplified, as shown in the above equation. It can be seen that the tunneling probability exponentially changes with the gap (insulating barrier width) between the conductive surface and the tip. As a result, any small change in the surface height translates to a measurable signal change by an ammeter.

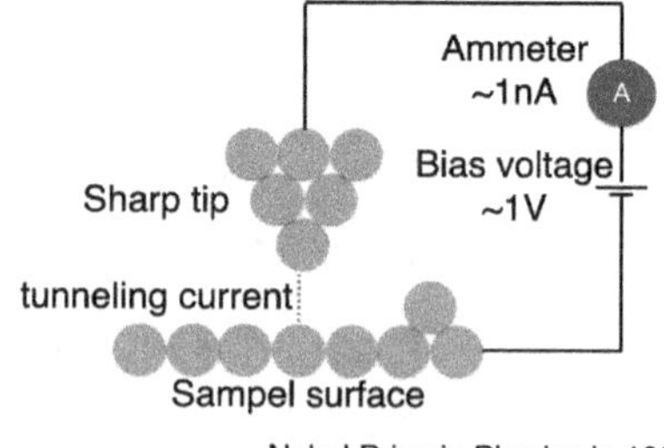

Nobel Prize in Physics in 1986

The solution for tunneling probability found in the previous section can be simplified in the regime where $\lambda a \ll 1$, as

$$T \simeq \frac{16\lambda^2\mu^2}{(\mu^2 + \lambda^2)^2} e^{-2\lambda a}.$$

The bias voltage, V, controls the kinetic energy of the electrons. The potential energy is given by the well-characterized vacuum gap. A typical value of $U - E = 5$ eV (electron volts) results in a decay constant of $\lambda \simeq 10^{10}$ or $1/A^\circ$. The current decays by one order of magnitude per $1\,A^\circ$.

It suggests that for every 0.1 nm of change in the height of the surface, the current magnitude changes by approximately an order of magnitude. This is the reason that STMs can map the surface roughness of a conductive surface with a resolution better than the size of a single atom. If the height of the surface changes by a single atom, an STM will be able to accurately detect the change by measuring change in the surface current.

1.3. Dirac Notation

In this section, we introduce a mathematical notation, known as Dirac notation, introduced by Paul Dirac in 1926. The notation enables us to write quantum equations in more compact forms.

Dirac (1926) introduced a very convenient notation to express wave functions in position space. In Dirac notation, a wave function in position space is represented by a "ket" or its complex conjugate by a "bra", where

ket
$\psi \equiv |\psi\rangle$ represents a column vector or a function describing the wave function.

bra
$\psi^* \equiv \langle\psi|$ represents a row vector or a complex conjugate of function describing the wave function.

Although this notation does not simplify the mathematics and quantum calculations, it does condense the formulation and makes it easier to write the equations. We shall see that "ket" and "bra" are vector representations of a wave function in a finite-dimensional space.

1.3.1. *Wavefunction Properties in Dirac Notation*

To express the integration using Dirac notation, we use "bra–ket" product:

$$\langle\psi|\phi\rangle = \int \psi^*\phi\, d^3r$$

For two wavefunctions which do not overlap in space (orthogonal), we can write

$$\langle\psi|\phi\rangle = \int \psi^*\phi\, d^3r = 0$$

To express expected value corresponding to an operator, we can write

$$\langle\psi|\hat{O}|\psi\rangle = \int \psi^*\hat{O}\psi\, d^3r$$

As seen above, instead of writing the full integral overlap of two wave functions ψ and ϕ, the Dirac notation enables us to write the integral in compact form using "bar" of ψ multiplied by "ket" of ϕ. If the two wave functions do not overlap, i.e. they are orthogonal, the result of "bra–ket" will be zero.

The form of the expectation value corresponding to the operator $\hat{O}$ can also be simplified in Dirac notation as seen above. To calculate the expectation value, we first apply the operator to the wave function and then multiply the result by the complex conjugate of the wave function. This notation allows us to keep track of the order of the operations. A few more properties of Dirac notation are shown below.

Inner products:

$$\langle\psi|\psi\rangle \qquad\qquad \text{Probability}$$
$$\langle\psi|\phi\rangle = 0 \rightarrow \psi \perp \phi \qquad\qquad \text{Orthogonal states}$$
$$\langle\psi|\phi\rangle = \langle\phi|\psi\rangle^* \qquad\qquad \text{Complex conjugate}$$
$$\langle\psi|c_1\phi_1 + c_2\phi_2\rangle = c_1\langle\psi|\phi_1\rangle + c_2\langle\psi|\phi_2\rangle \qquad\qquad \text{Superposition}$$

Note that $\langle\psi|\psi\rangle$ is the probability of finding a particle whose wave function is $|\psi\rangle$ in a region of space, which is calculated by taking the integral of the probability density, $\psi^*\psi$, over that region of space.

If the two wave functions are orthogonal, the overlap integral is zero. Conversely, if the two wave functions perfectly overlap, the total probability or integral overlap equals 1.

To calculate the complex conjugate of "bra" or "ket", we exchange "bra" and "ket".

If the wave function ϕ is written in terms of the superposition of two "sub"-wave functions (basis functions), ϕ_1 and ϕ_2, to calculate $\langle \psi|\phi \rangle$, we can write it as a linear sum of $\langle \psi|\phi_1 \rangle$ and $\langle \psi|\phi_2 \rangle$. This property can be simply proven using the integral form of $\langle \psi|\phi \rangle$.

Now, let us consider again the example of a particle in a harmonic potential, where the wave function of the particle can be written in terms of the eigenfunctions of the Hamiltonian operator. The eigenfunctions form the orthonormal basis of the function space on which the Hamiltonian operator is defined.

We saw that the solution for a harmonic potential is of the form

$$\phi_n(x) = u_n(x)e^{-m\omega x^2/2\hbar}$$

n	$u_n(x)$
0	$\left(\frac{1}{a\sqrt{\pi}}\right)^{1/2}$
1	$\left(\frac{1}{2a\sqrt{\pi}}\right)^{1/2} 2(x/a)$
2	$\left(\frac{1}{8a\sqrt{\pi}}\right)^{1/2} [2 - 4(x/a)^2]$
3	$\left(\frac{1}{48a\sqrt{\pi}}\right)^{1/2} [12(x/a) - 8(x/a)^3]$
$\vdots$	
n	$\left(\frac{1}{n!2^n a\sqrt{\pi}}\right)^{1/2} H_n(x/a)$

$$\langle \phi_{n=0}|\phi_{m=1} \rangle = \int_{-\infty}^{\infty} \underbrace{\frac{1}{\sqrt{a\sqrt{\pi}}}e^{-m\omega x^2/(2\hbar)}}_{u_0(x)} \cdot \underbrace{\frac{1}{\sqrt{2a\sqrt{\pi}}}2\frac{x}{a}e^{-m\omega x^2/(2\hbar)}}_{u_1(x)}\, dx$$

where the braces indicate $\psi_0^*(x)$ and $\psi_1(x)$ respectively.

$$\longrightarrow \int_{-\infty}^{\infty} xe^{-m\omega x^2/\hbar}\, dx = 0.$$

We note again that the length constant $a = \sqrt{\hbar/(m\omega)}$. For a particle in a single energy state, the quantum number n identifies the Hermit polynomial function which describes the position dependency of the wave function.

One can see that the integral overlap between any two eigenfunctions with different quantum numbers is zero. This allows us to write the product of two eigenfunctions as a Dirac delta function (whose value is zero everywhere except at zero).

The wave function of the particle is often more complex, consisting of multiple energy states written in a coherent superposition form. We can think of $|\phi_n\rangle$ as a set of orthogonal vectors with unity length, constructing a coordinate system, which enables us to describe the energy of the particle in a multi-dimensional space. This multidimensional space is called the Hilbert space. For example, a particle in a superposition of the first three energy states, $c_0|\phi_0\rangle + c_1|\phi_1\rangle + c_2|\phi_2\rangle$, can be considered as a particle in a three-dimensional space where $c_{0,1,2}$ describes the contribution of each orthogonal axis (energy) in defining the particle's energy. In this case, c_0, c_1, and c_2 form a set of basis vectors for our three-dimensional Hilbert space.

1.3.2. *Vectorial Representation of Wave Functions*

To see the connection between the Dirac notation and the mathematical formalism in the classical picture, consider the sinusoidal steady-state analysis performed in electronics. In classical electronics, the phasor form of a voltage waveform, which is a single-frequency wave represented by a vector, is used to simplify the math. In this representation, a vector in the complex plane is used to describe the amplitude and phase of the waveform.

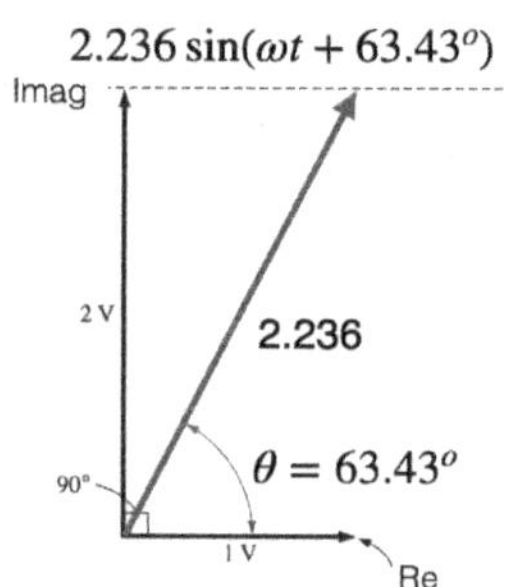

We use vectors as phasor form of a wave in classical electronics. In the example above, a sinusoidal oscillation of current or voltage can be illustrated by a vector whose length is equal to the amplitude of the oscillation and its angle from the real axis (Re) is the phase of the signal. We omit the oscillation frequency in this representation considering a frame rotating at frequency ω.

Similarly, in finite dimensional QM space, wave functions can also be defined as vectors using

$$\langle U| = (U_1^* \ U_2^* \ \dots \ U_N^*) \quad |V\rangle = \begin{pmatrix} V_1 \\ V_2 \\ \vdots \\ V_N \end{pmatrix}.$$

So, the inner product can be written as a matrix multiplication of a row vector with a column vector.

Such vector representations can be further simplified by the Dirac notation where a "bra" is a row vector and a "ket" is a column vector. This simplification enables us to calculate the result of operations or integral on the wave functions using the matrix multiplication formalism.

1.3.3. *Basis Functions*

We can describe an arbitrary vector in a space using the unit vectors of that space. In quantum mechanics, this is called a basis of the space in which an operator is defined. For instance, consider a two-dimensional vector $\langle \psi| = (a^* \ b^*)$. We can write this vector (or the wave function) as $\langle \psi| = a^*(1\ 0) + b^*(0\ 1)$ or $|\psi\rangle = a\begin{pmatrix} 1 \\ 0 \end{pmatrix} + b\begin{pmatrix} 0 \\ 1 \end{pmatrix}$.

$$|\psi\rangle = \sum_n c_n |\phi_n\rangle$$

The wave function of a particle can be written in terms of its basis functions:

$$|\psi\rangle = \begin{pmatrix} a \\ b \end{pmatrix} = a\begin{pmatrix} 1 \\ 0 \end{pmatrix} + b\begin{pmatrix} 0 \\ 1 \end{pmatrix}.$$

In higher dimensions, more unit vectors or basis functions are needed to fully describe the wave function. For example, a particle in a harmonic potential can be described in terms of energy basis functions

$$\phi_n(r, t) \equiv |\phi_n\rangle \equiv |n\rangle.$$

Basis functions are orthonormal, i.e.

$$\langle \phi_m | \phi_n \rangle \equiv \langle m | n \rangle = \delta_{mn}.$$

The unit vectors or basis functions are orthonormal and any wave function can be written as a superposition of its basis functions as shown in the figure above. A particle with harmonic potential, for example, can be written in terms of its energy basis functions or eigenstates of the Hamiltonian operator. Note that in some cases, we use only the quantum number n to represent the Hamiltonian eigenstates as $|n\rangle$. The basis functions are orthonormal, i.e. the products of two energy eigenstates is a Dirac delta function, $\delta_{mn} = 1$ (when $m = n$) or $\delta_{mn} = 0$ (when $m \neq n$).

Depending on what property of the wave function we like to study, we write the wave function in terms of the corresponding basis functions. For example, if we consider an electromagnetic field, its wave function can be described in different basis functions, e.g. polarization, frequency, time, and photon number. If we want to understand how the polarization changes as it interacts with matter, we can express the wave function using the polarization basis and ignore other degrees of freedom. This assumption is valid as long as the interaction changing the polarization does not affect other properties of the photon important to us. Otherwise, we need to write the wave function as a product of multiple basis functions, each describing the dynamics of a different degree of freedom of the particle.

Consider a wave function written in terms of its complete basis functions using Dirac notation:

$$|\psi\rangle = \sum_i c_i |i\rangle \quad \langle\psi| = \sum_i c_i^* \langle i|$$

We can calculate the probability by taking its integral overlap written using the Dirac notation:

$$\langle\psi|\psi\rangle = \left(\sum_i c_i^*\langle i|\right)\left(\sum_j c_j|j\rangle\right) = \sum_{i,j} c_i^* c_j \delta_{ij} = \sum_i c_i^* c_i = 1$$

We write the wave function in terms of its complete basis functions, represented by $|i\rangle$. The discrete value of i can refer to oscillation in a specific directions or the quantized energy of a particle. To write the complex conjugate of the wave function in the Dirac notation, we change "ket" to "bra" and coefficients c_i to c_i^*.

The wave function needs to be normalized. By calculating the "bra–ket" and using the fact that the eigenstates $|i\rangle$ are orthonormal, i.e. $\langle i|j\rangle = \delta_{ij}$, we find that sum of square of absolute values of probability amplitudes should be equal to one.

1.3.4. *Operators in Dirac Notation*

We can also write an operator in a finite-dimensional space using the matrix formalism. In this case, each matrix element of the operator can be expressed in terms of its basis functions. Below, we provide a simple proof that if an operator's matrix element is defined as $O_{ij} = \langle i|\hat{O}|j\rangle$, then the matrix form of the operator is given by $\sum_{i,j} O_{ij}|i\rangle\langle j| \equiv \hat{O}$.

We can verify that operators can be written as matrix elements such that

$$O_{ij}|i\rangle\langle j| \equiv \hat{O} \quad \text{if } O_{ij} = \langle i|\hat{O}|j\rangle.$$

To show this, we can multiply both sides by $\langle i|$ from left and $|j\rangle$ from right:

$$\langle i|\cdot O_{ij}|i\rangle\langle j|\cdot|j\rangle = \langle i|\cdot\hat{O}\cdot|j\rangle$$

$$O_{ij}\langle i|i\rangle\langle j|j\rangle = \langle i|\hat{O}|j\rangle$$

$$O_{ij} = \langle i|\hat{O}|j\rangle$$

Note that O_{ij} is a number and not an operator, and can be taken out of the integral. Knowing that $|j\rangle$ and $|i\rangle$ are orthonormal, we arrive at the conclusion that an operator's matrix elements, O_{ij}, can be calculated by finding the integral overlap $\langle i|\hat{O}|j\rangle$.

The identity operator is the simplest operator, it can be written in terms of the basis functions of the state. For example, $\hat{I} = \sum_n |n\rangle\langle n| = \sum_n |\phi_n\rangle\langle\phi_n|$ is the identity operator, where $|n\rangle$ or $|\phi_n\rangle$ are the basis functions. In this case, $I_{ij} = \delta_{ij}$. When the identity operator is applied to the wave function, the resulting wave function is unchanged.

1.3.5. *Example: Projective Measurement*

Let us now consider another example where a quantum particle with quantized energy oscillates and shows wave characteristics. The oscillation can be in two directions x and y (a two-dimensional space) and the wave function can be described as a superposition of oscillations along the x and y directions. The particle's wave function can then be written in terms of its basis functions $|\phi_x\rangle$ and $|\phi_y\rangle$.

We consider the particle initially prepared in state

$$|\Psi\rangle = c_x|\phi_x\rangle + c_y|\phi_y\rangle = \begin{pmatrix} c_x \\ c_y \end{pmatrix}$$

where $|\phi_{x,y}\rangle$ are orthogonal (non-overlapping) basis functions of a particle representing oscillations along the x or y directions.

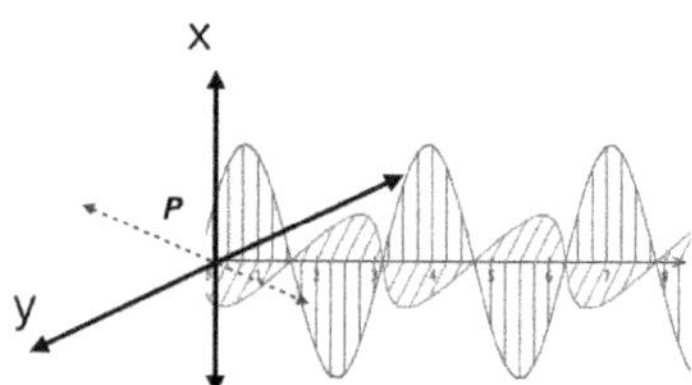

The vector representation of the wave function is also provided above in the first equation.

A quantum particle corresponding to the electromagnetic field (i.e. a photon) is a more specific example that can possess such oscillations. A photon with polarization along vertical, V, or horizontal, H, directions or any combination of these directions is a quantized particle that shows oscillations along those directions. Assuming that the energy of the system remains constant, we can write the wave function of the photon by only considering the polarization or direction of the field's oscillations. Oscillations along any direction can be represented by coherent suppression of different eigenstates of the polarization operator (i.e. $|\phi_y\rangle = |H\rangle$ and $|\phi_x\rangle = |V\rangle$) that form the complete basis.

Now, consider a detector that is designed to detect the oscillations or the polarization of a photon. If the detector is aligned along the vertical axis, it will register a single click when a photon is vertically polarized. Conversely, a detector aligned along the horizontal direction registers a click only when the photon is horizontally polarized.

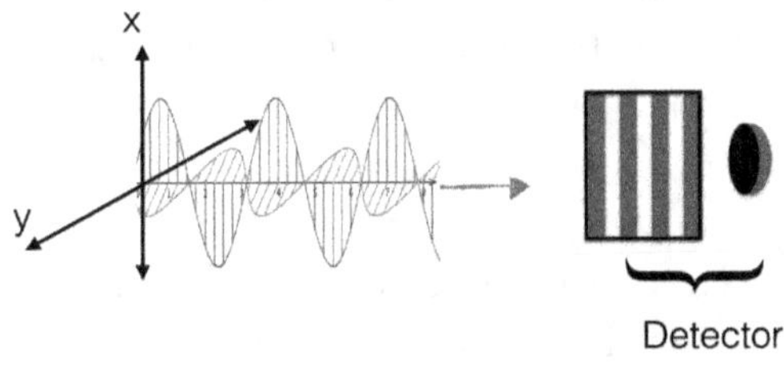

$$|\Psi\rangle = c_x|\phi_x\rangle + c_y|\phi_y\rangle$$

For the wave function of the form above, we can show that the detector will click with probability

$$|\langle\phi_x|\Psi\rangle|^2 = |c_x|^2.$$

Such measurement causes the wave function to **collapse** to state $|\phi_x\rangle$.

To describe the measurement operation, we define the term "projective measurement". When the photon or particle is in a superposition of two oscillation directions, the detector registers x or y oscillation with some probability. For example, to find the probability of detecting the particle oscillating along the direction x, we calculate the square of the projection amplitude shown above. This suggests that if we repeat the experiment many times, $|c_x|^2$ fraction of times, the detector will register a particle with oscillation along the x direction. The projection operator in this case is then given by $|\phi_x\rangle\langle\phi_x|$ and its expectation value is $\langle\psi|(|\phi_x\rangle\langle\phi_x|)|\psi\rangle = |c_x|^2$.

Whenever such measurement occurs, we can state that the particle's wave function has collapsed to the state $|\phi_x\rangle$. In this basis, the wave function no longer has a quantum mechanical description such as a superposition state, in this basis. We shall see later that when the wave function consists of more than one particle, the projective measurement of one particle may change the state of the other particles (in the case of entanglement).

1.4. Density Operator

In this section, we introduce the density operator and its matrix form, the density matrix. To find a proper model for many experiments, the density operator is frequently used as opposed to the wave function. This is because, in practice, a single wave function cannot fully describe the system. When a system can be described by a single wave function, the state of the system is called a pure state. However, due to interactions with the environment (thermal bath), the system is often found in a statistical mixture of different wave functions. In this case, the system can be described as a mixed state.

So, a wave function is one way to describe a quantum state:

$$|\Psi\rangle$$

Alternatively, a quantum state can be described by its density matrix.

The density matrix is defined as outer product of the wave function and its conjugate:

$$\hat{\rho} = |\Psi\rangle\langle\Psi|.$$

For a not pure state (mixed state), we have

$$\hat{\rho} = \sum_i p_i|\Psi_i\rangle\langle\Psi_i|.$$

A density operator is formed by the outer product of the wave function with its complex conjugate. So, the density operator contains the same information as the wave function and can also be used to describe the quantum state.

A density operator captures the dynamics of a mixed state, which leads to a more accurate model for many physical systems. We note that a statistical mixture is different from a superposition. A quantum state that is in a superposition state of two different wave vectors, i.e. $|\psi\rangle = c_1|\phi_1\rangle + c_2|\phi_2\rangle$, has a density operator with terms like $|\phi_1\rangle\langle\phi_2|$ signifying the interference (or coherence) between states $|\phi_1\rangle$ and $|\phi_2\rangle$. A mixed state, however, cannot be written with a single wave function, and the density operator, therefore, allows us to capture the nature of the statistical mixture as well as the superposition and coherence nature of the state. A density operator written as above for a system in statistical mixture of $|\psi_1\rangle$ and $|\psi_2\rangle$ is written as $\hat{\rho} = \sum_i p_i |\psi_i\rangle\langle\psi_i|$. We can see that this density matrix does not have any term suggesting interference between $|\psi_1\rangle$ and $|\psi_2\rangle$, so it describes a mixed state, i.e. it is not a pure quantum state.

When the quantum state is written as a superposition of its basis functions, the density operator of such a pure state can be found by taking the outer product of the state with its complex conjugate.

Given a wave function $|\Psi\rangle$, we can write the density operator in terms of the density matrix elements

$$|\psi\rangle = \sum_i c_i |\phi_i\rangle \longrightarrow \text{where } \hat{\rho} = \sum_{ij} c_i c_j^* |\phi_i\rangle\langle\phi_j|$$

where $\hat{\rho} = |\psi\rangle\langle\psi|$ is the density operator and density matrix elements are

$$\rho_{ij} = \langle\phi_i|\hat{\rho}|\phi_j\rangle = c_i c_j^*$$

where $i = j$ denotes population/probability of finding particle in state $|\phi_i\rangle$ and where $i \neq j$ denotes coherence or "waveness" of a particle.

As can be seen above, the density operator is written as a sum of the outer products of the basis functions. The coefficients $c_i c_i^*$ describe the probability of finding the system in each state $|\phi_i\rangle$ and we refer to them as the population. The coefficients $c_i c_j^*$ $(i \neq j)$ are coherence terms that describe the interference and superposition of different wave vectors.

Using the Dirac notation, the density operator in a finite-dimensional space can be expressed as a matrix with diagonal terms being the populations and the off-diagonal terms being the coherences. The density matrix elements can be measured experimentally (assuming a finite-dimensional space) to evaluate the degree of "quantumness" of the system. If the density matrix contains only diagonal elements, the system is in a mixed state. Mathematically, the trace of $\hat{\rho}^2$ is equal to 1 for a pure state and less than 1 for a mixed state. The diagonal elements can simply be measured by finding the average number of times the system is found in a particular state of certain, for example, energy, polarization, or frequency.

1.4.1. *Example: Density Matrix of a Qubit*

Now, consider the example from earlier; an oscillating particle is in a superposition of two oscillation directions, x and y. Suppose we have a detector that can detect the particle oscillating along a specific direction.

As before we consider a particle initially prepared in state:

$$|\Psi\rangle = c_x|\phi_x\rangle + c_y|\phi_y\rangle$$

where $|\phi_{x,y}\rangle$ represents the orthogonal (non-overlapping) wave functions of a particle oscillating along the x or y directions.

Suppose we have a detector which can be set to measure particles oscillating along the x, y or diagonal direction.

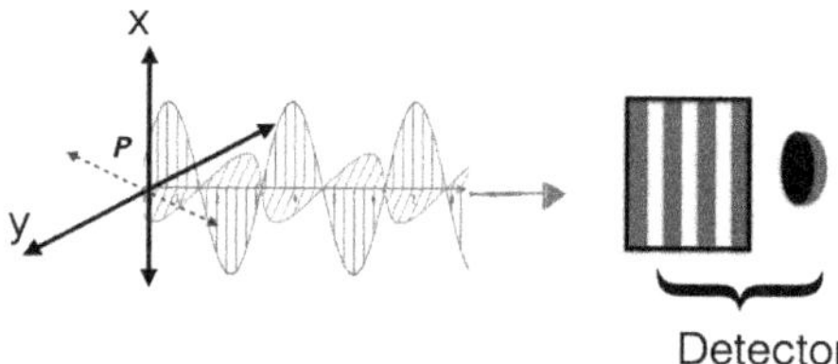

The detector can be set to measure the x, y, or an arbitrary direction P. This sets the basis of the measurements. In the case of photonic quantum particles, the oscillation can refer to the polarization where a polarization-sensitive detection can be achieved by using polarization optics to transmit a specific polarization followed by single-photon detectors.

The density operator corresponding to the pure state of the particle is given below.

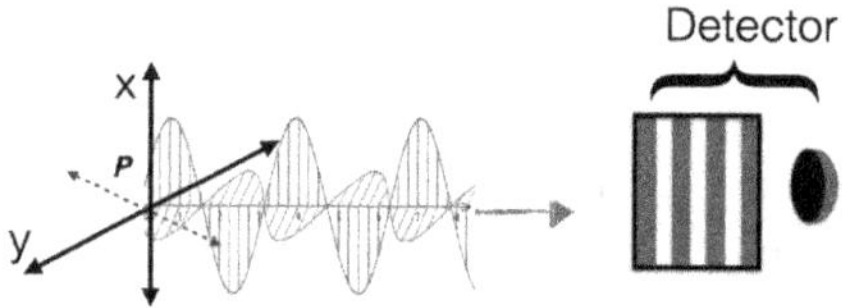

$$\hat{\rho} = p_{xx}|\phi_x\rangle\langle\phi_x| + p_{yy}|\phi_y\rangle\langle\phi_y| + p_{xy}|\phi_x\rangle\langle\phi_y| + p_{yx}|\phi_y\rangle\langle\phi_x|$$

- p_{xx} and p_{yy} can be measured by averaging clicks when detector is oriented along the x or y direction.
- p_{yx} and p_{xy} can be computed by measuring the visibility of interference between the two oscillating states by setting the detector at 45 degrees from the x axis.

The coefficients p_{ij} describe the population and coherence terms of the density matrix. To experimentally measure p_{xx} and p_{yy}, the detector needs to be set such that it can detect a particle with oscillations along the x or y direction. By repeating the measurement in these bases, the probability of detection in each basis is then identified.

For measurement of the coherence terms, p_{xy} and p_{yx}, the detector is aligned at ± 45 degrees from the x direction to detect the x and y oscillations with equal contributions. By repeating this measurement and applying various delays between the x and y oscillations before detection, one can obtain an interference fringe. The visibility of the interference fringe gives information about the coherence terms. In the case of photonic qubits, for example, the relative delay between V and H polarization can be achieved using birefringent crystals (e.g. quarter-wave plates), which apply a different delay between a V-polarized photon compared to an H-polarized photon.

The density matrix in this example consists of p_{ij} elements.

$$\rho_{ij} = \begin{pmatrix} \cos^2 \theta & \cos \theta \sin \theta e^{i\phi(t)} \\ \cos \theta \sin \theta e^{-i\phi(t)} & \sin^2 \theta \end{pmatrix}$$

It is a 2×2 matrix with the diagonal elements taking the form of sin and cos of the polarization measurement angle with respect to the vertical and horizontal axes. The off-diagonal elements are the coherence terms determined by the visibility measurement when the two polarization states interfere.

In the case of quantum optical states in the polarization basis, the coherence terms can be understood by considering the phase noise between polarization axes. If a photon in a superposition of two polarizations, V and H, propagates through a medium with large fluctuating refractive index, the coherence terms can vanish. This can happen, for example, when a fiber channel undergoes temperature or stress fluctuations. As a result, the phase term, $\phi(t)$, becomes time-dependent and its value randomly changes over time. Therefore, repeating the measurements to construct a statistical distribution of measurement results involves $\sum_t e^{i\phi(t)}$ which asymptotes to zero when fluctuations are large. This causes the off-diagonal elements or coherence terms that are representative of "quantumnees" in the system to vanish.

1.5. Commutation Relation

In this section, we introduce commutation relationship between operators and its connection with Heisenberg Uncertainty Principle.

Two operators $\hat{A}$ and $\hat{B}$ commute if

$$[\hat{A}, \hat{B}] = \hat{A}\hat{B} - \hat{B}\hat{A} = 0$$

When two operators do (not) commute, it is (not) possible to know their respective eigenvalues with complete accuracy. In that case, we write the following inequality (i.e. Uncertainty Principle):

$$\Delta A \Delta B \geq \left| \frac{i}{2} \langle [\hat{A}, \hat{B}] \rangle \right|$$

The notation $[\hat{A}, \hat{B}]$ indicates the commutation between the operators. This expression quantifies whether exchanging the operators' order affects the final result when they are both applied to a quantum state.

To understand the meaning of commutation relations, consider this classical analogy. Think of a blank piece of paper. One can perform different "operations" on the paper. For example, painting with different orders of colors used could be thought of as applying different operations. If this is a digital painting, one could also consider applying additional operations by digitally processing the image (e.g. smoothing). Depending on the type of "operations" we perform, the result may or may not depend on the order of the operations.

If the only two operations performed on the paper are drawing the top half of the paper and the bottom half of the paper, the order in which these operations are performed, ideally, does not affect the final results. However, if the two operations are drawing and smoothing, the final result depends on the order of the operations.

When the two quantum operators, $\hat{A}$ and $\hat{B}$, do not commute, i.e. $[\hat{A}, \hat{B}] \neq 0$, there is a minimum uncertainty associated with performing both operations, $\Delta\hat{A}$ and $\Delta\hat{B}$. The relationship between the uncertainty and the commutation relation given above is known as the Uncertainty Principle, suggesting the precise determination of non-commuting observable is not possible. Here, the notation $\langle \cdots \rangle$ indicates the expectation value of the commutation. Note that $\Delta\hat{A}$ and $\Delta\hat{B}$ are not operators but measurable quantities.

Two of the most commonly used operators in quantum mechanics are position and momentum operators. To find whether two typical operators $\hat{x}$ and $\hat{p}_x$ commute or not, one needs to apply the commutation operator to the wave function of interest.

$$[\hat{x}, \hat{p}_x]\psi = -i\hbar \left(x\frac{\partial\psi}{\partial x} - \frac{\partial(x\psi)}{\partial x} \right)$$

$$[\hat{x}, \hat{p}_x] = i\hbar$$

$$\Delta A\Delta B \geq \left| \frac{i}{2}\langle [\hat{A}, \hat{B}]\rangle \right| \rightarrow (\Delta x)^2(\Delta p_x)^2 \geq |i\hbar|^2/4$$

$$\rightarrow \Delta x\Delta p \geq \left| \frac{i}{2}\langle [\hat{x}, \hat{p}]\rangle \right| \rightarrow \Delta x\Delta p_x \geq \hbar/2$$

In the above example, we see that when we use the differential form of the momentum operator and take derivatives of $\psi(x)$ and $x\psi(x)$, we find that the result of applying the commutation to $\psi(x)$ is $i\hbar\psi(x)$. Therefore, we can write the commutation relation between $\hat{x}$ and

$\hat{p}_x$ as shown by the second equation above. These operators do not commute. Therefore, there must be an uncertainty relation associated with them when both operations are performed. In other words, when we measure the expectation values associated with both operators simultaneously, the product of uncertainty in measuring both values has a lower bound.

The minimum uncertainty is directly related to the commutation relation between the operators. We can then find that the standard deviation in measuring the position and momentum of a particle in a 1D space has a minimum of $\hbar/2$.

This does not mean that one cannot measure the position or momentum of the particle with 100% certainty. It indicates that one cannot measure simultaneously **both** position and momentum with 100% certainty. The more precisely you measure one, the less precisely you can measure the other non-commuting quantity.

1.5.1. *Quantum Microscope*

Let us consider a "quantum microscope" example, where observation of a quantum particle, such as an electron or an atom, is the goal. If we use the electromagnetic field or photons to detect the particle, we see that there is a minimum uncertainty in measuring both position and momentum of the particle.

Suppose we have a quantum microscope and like to measure position and speed (momentum) of an electron.

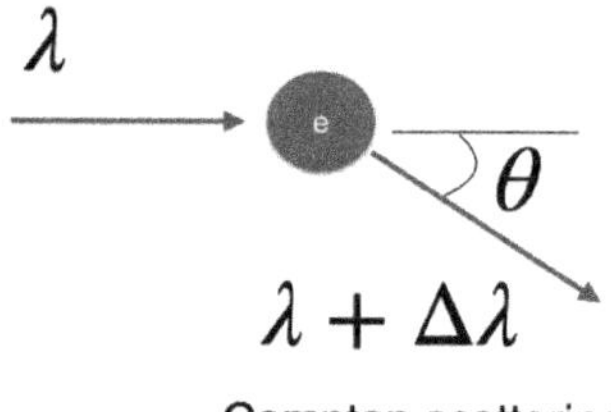

Compton scattering

A photon used to detect the electron will be scattered, transferring some of its momentum to the electron:

$$\Delta p_{\min} = \frac{h\Delta\lambda}{\lambda^2}$$

Minimum resolvable distance is given by the diffraction limit as

$$\Delta x = \frac{\lambda}{\underbrace{2n\sin\theta}_{\mathrm{NA}}} \simeq \frac{\lambda}{2.8}.$$

Let us discuss these uncertainties further. When the incident photon inelastically scatters off the particle, its frequency (or wavelength) will change. This can be understood classically using the Doppler effect when the particle has a non-zero velocity. Consequently, the momentum of the particle also changes. As the momentum of the particle is $p = \hbar k$, where $k = 2\pi/\lambda$, the minimum change in the momentum, $\Delta p_{\min}$, and change in the wavelength, $\Delta\lambda$, can be related using the equation above. The minimum resolvable size of the particle, Δx, in the direction perpendicular to the incident photon propagation is also given by the projection of δx to a plane perpendicular to the scattered photon. One can increase the numerical aperture (NA) of the microscope, $\mathrm{NA} = 2n\sin\theta$, where n is the refractive index of the medium, to increase the resolution in determining Δx. The best resolution of the optical microscope is typically on the order of $\lambda/2.8$, which is referred to as the diffraction limit. Although the treatment above is not completely quantum mechanical, one can use the analogy to see that the uncertainty in measuring momentum and uncertainty in measuring the position of a particle have opposite relationships with wavelength λ. The larger the wavelength of the photon, the smaller the momentum uncertainty and the larger the position uncertainty.

1.5.2. *Calculation of Uncertainty*

To explicitly calculate the error in measuring an observable, we need to calculate the expectation values:

$$\Delta x = \sqrt{\langle x^2 \rangle - \langle x \rangle^2} \qquad \Delta p = \sqrt{\langle p^2 \rangle - \langle p \rangle^2}.$$

Using quantum theory, for example, we have

$$\langle p^2 \rangle \equiv \langle \Psi | \hat{p}^2 | \psi \rangle = \int_{-\infty}^{\infty} \psi(x)^* \left(-\hbar^2 \frac{d^2}{dx^2} \right) \psi(x)dx.$$

Using the operator language and the expectation values of the operators, one can more rigorously calculate the uncertainty relationship between the two operators. Consider again the position and momentum operators. The approach is similar to finding standard deviation in classical statistics. If the wave function of the system or particle is known, one can calculate the expectation values for $\hat{x}$ and $\hat{x}^2$ operators (similarly for $\hat{p}$) to calculate the uncertainty in position and momentum using the first two equations above. This enables us to estimate the uncertainty in measuring each observable.

Note that explicitly calculating the expectation values requires taking multiple integrals, one of which is shown in the last equation above. To take such an integral, the form of the operator and wave function should be provided. Note that given certain wave functions, Δx or Δp can be zero when measured independently. To find the minimum uncertainty when both x and p quantities are measured simultaneously, one needs to use the Heisenberg Uncertainty Principle.

1.5.3. *Fourier Transform and Uncertainty Relation*

Let us now consider another semi-classical example that may allow us to see the minimum uncertainty through a different lens. The electromagnetic field of a single photon is a spatial–temporal oscillation of the electric and magnetic fields. For now, we will consider an electric field which has a Gaussian temporal intensity profile and oscillates with frequency ω, while propagating along the z direction.

$$k = \frac{2\pi}{\lambda} = \frac{\omega}{c} = \frac{p}{\hbar}$$

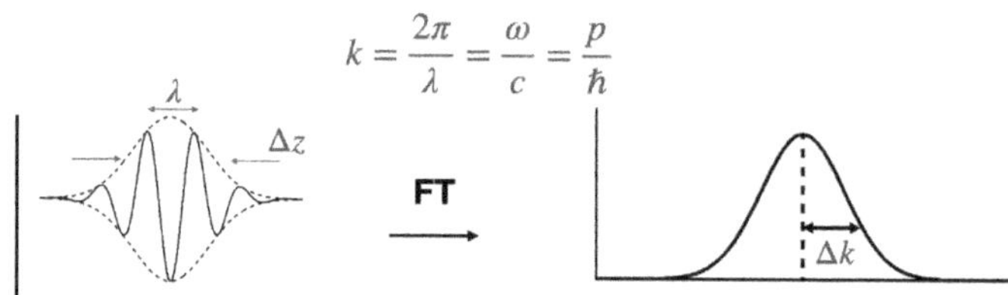

An EM field with Gaussian profile in space can be written as

$$E(z,t) = E_0 e^{-\frac{(ct-z)^2}{2\sigma_z^2}} \cos(\omega t - kz)$$

$$\Delta z = \frac{\sigma_z}{\sqrt{2}}.$$

The quantity k is the wave number. The dashed envelope of the field on the left-hand side is the wave packet of the field propagating with a certain group velocity in the vacuum. The oscillations within the envelope have a frequency ω and a phase velocity c. Since the photon pulse or the wave packet has a temporal duration, it occupies some region in space, σ_z. This is to say that the photon as a particle can be found at location $z - ct$ at time t with uncertainty Δz. One can study the photon in the Fourier plane by taking spatial Fourier transform (FT) to find equivalent minimum uncertainty in a photon's wave number or momentum, as shown above:

$$\Delta k = \frac{1}{\sqrt{2}\sigma_z}$$

$$\Delta p = \frac{\hbar}{\sqrt{2}\sigma_z}.$$

Multiplying these two uncertainties gives the minimum uncertainty of position and momentum along the z direction of the photon's wave packet:

$$\Delta k \Delta z = \frac{1}{2}$$

$$\Delta p \Delta z = \frac{\hbar}{2}.$$

Let us consider the same EM field with Gaussian profile in space. We can translate the uncertainty in position to uuncertainty in time

$$E(z,t) = E_0 e^{-\frac{(ct-z)^2}{2\sigma_z^2}} \cos(\omega t - kz)$$

$$\Delta z = \frac{\sigma_z}{\sqrt{2}}, \quad \Delta t = \frac{\Delta z}{c} = \frac{\sigma_z}{c\sqrt{2}}.$$

By considering the uncertainty in time instead of position, we can derive an uncertainty principle for energy and time for a photon wavepacket. The temporal duration of the E-field is given by Δt, as shown above. In the particle picture, the photon centered at time t will arrive within a window of $\pm \Delta t$. We can also take the temporal Fourier transform to obtain the spectral decomposition of the photon and its width, which is given by $\Delta\omega$,

$$\Delta\omega = c\Delta k = \frac{c}{\sqrt{2}\sigma_z}.$$

Multiplying the minimum uncertainties associated with measuring the frequency and arrival time of the photon yields another uncertainty relationship:

$$\Delta\omega\Delta t \geq \frac{1}{2}$$

$$\Delta E\Delta t \geq \frac{\hbar}{2}.$$

As we shall see later, photons can be considered particles in a harmonic potential with quantized energy, $\hbar\omega$. So, multiplying the $\Delta\omega$ by $\hbar$ gives the energy uncertainty, ΔE, which enables us to write the energy–time uncertainty relation for $\Delta E\Delta t$.

This relation indicates that accurately measuring both the time and the frequency (or the energy) of a photon is not possible. This uncertainty in energy can also be linked to uncertainty in the frequency of the atom (line width of the atom) emitting the photon. In other words, although the energies of quantum particles such as atoms and photons are quantized, their energies cannot be accurately determined at a specific time t.

1.6. Entanglement

Entanglement is a core concept in quantum mechanics. A concept that sets apart quantum systems from their classical counterparts. Entanglement is a superposition extended to more than one particle separated in space. When two or more particles are in an entangled state, they are collectively described by a single wave function capturing the phase relationship (coherence) between the non-local particles. In the case of computational qubits, entanglement between multiple qubits enables "super-parallel" computation, resulting in an exponential speed-up in computation, at least in principle. For communication qubits such as photons, entanglement enables secure communication using quantum teleportation. Once two entangled photons are shared between two parties, e.g. Alice and Bob, one of the entangled photons at Alice's location can be interfered with a quantum state carrying information to be teleported to the other twin photon at Bob's location. Joint detection of states by Alice followed by classical communication of measurement results to Bob teleports quantum information without directly sending a quantum state from Alice to Bob. Other applications of entanglement in quantum communications include distributed quantum computing where a gate operation is teleported to connect multiple quantum processors, making a more powerful processor.

In this section, we introduce the concept of entanglement and discuss how entangled photons can be created probabilistically. We note that there are ways to create entangled photons with high certainty (deterministically), which we will discuss later in the following chapters.

1.6.1. *The Schrödinger's Cat*

Entanglement is one of the most counter-intuitive aspects of the quantum world.

Schrödinger coined the term 'entanglement' in his famous cat-paradox hypothetical experiment.

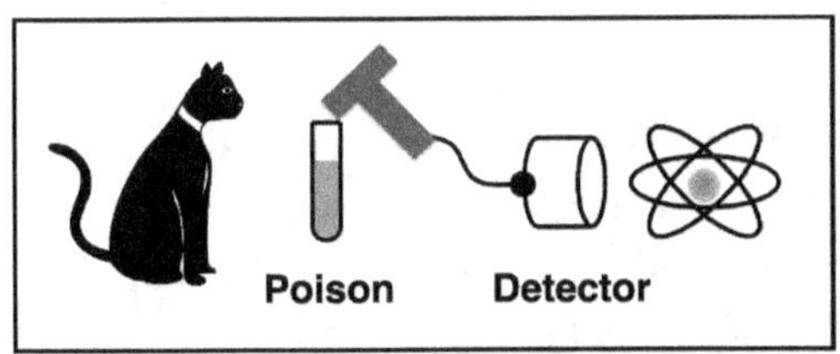

The famous thought experiment proposed by Schrödinger in a discussion with Einstein in 1935 is known as the Schrödinger's Cat experiment. Consider a scenario where a live cat, a glass containing poison, a particle detector, and a radioactive element are placed inside a box. The radioactive element will decay spontaneously causing the emission of a particle (e.g. gamma ray) at a random time. If the particle is detected by the detector (initially in the dark mode, $|1\rangle$), the state of the detector will change from $|1\rangle$ to $|2\rangle$. Once the detector clicks, it sends a signal to a device breaking the glass, poison is released and the cat dies. The total state of the system can be written as

$$|\Psi\rangle = \frac{1}{\sqrt{2}}(|\text{live}, 1\rangle + |\text{dead}, 2\rangle).$$

Of course, this is not a physical scenario and entanglement can not be obtained between the macroscopic cat and the atomistic radioactive element. In practice, as the relative phase between the $|1\rangle$ to $|2\rangle$ states randomly changes (due to the spontaneous nature of the decay process), one cannot write the state as a coherent superposition of $|1\rangle$ to $|2\rangle$ states. Nevertheless, a state of this form if can be obtained, where at least two elements of a system, in this case, the radioactive element and the cat, are coherently intertwined is an "entangled state". This thought experiment was initially proposed to highlight the problems with the Copenhagen interpretation of quantum mechanics.

1.6.2. *Quantum Entanglement and Correlation*

Multiple particles can be entangled if the wave function

$$|\Psi\rangle = |1_x\rangle|2_x\rangle + |1_y\rangle|2_y\rangle$$

describing the two particles cannot be factorized into a product of two wave functions.

- Entanglement is the amount of correlation between particles beyond what is classically possible.
- Note that entanglement is only defined for specific degrees of freedom (basis) of particles or waves.

To provide a mathematical definition of a state that is entangled, we refer to the quantum wave function. An entangled state of two particles is written as a coherent superposition of a particle's degrees of freedom. An entangled state can be written as the product of the states of the two particles. The example of the quantum state $|\Psi\rangle$ shown above describes two particles, 1 and 2, which are entangled in basis x and y. In other words, it is the x and y degrees of freedom of the two particles that are entangled. In this example, the x and y of particle 1 is correlated with x and y of particle 2.

Note that entanglement must involve at least two particles and the basis of entanglement should be identified.

To understand the concept of entanglement, let's use a classical analogy. Consider the Twin Family study done in 1993. If we plot the height of many twins over a period of time, we see a correlation between the height of twin one and twin two. The identical twins' heights are correlated. Instead of two people, let's now consider two quantum particles, e.g. two photons or two electrons, and instead of height let's consider correlation in the oscillation amplitude of the particles. We can then plot similar correlation curves for these "particle twins". The question is how to differentiate between classical correlations and quantum correlations in these examples. In both examples, the data points can be fitted with a linear line indicating correlations. However, the deviation of these data points (fluctuation) around the average is key to differentiating classical and quantum correlations. Quantum-correlated particles show fluctuations in correlation below what is fundamentally possible in classical systems. This fundamental limitation in fluctuation is related to the Heisenberg uncertainty limit. In the case of the twin study, the fluctuation physically originates from stochastic processes in gene mutation or chemical processes in biological systems. In the quantum case, the **relative** fluctuations can be suppressed by coherent and nonlinear interaction of particles.

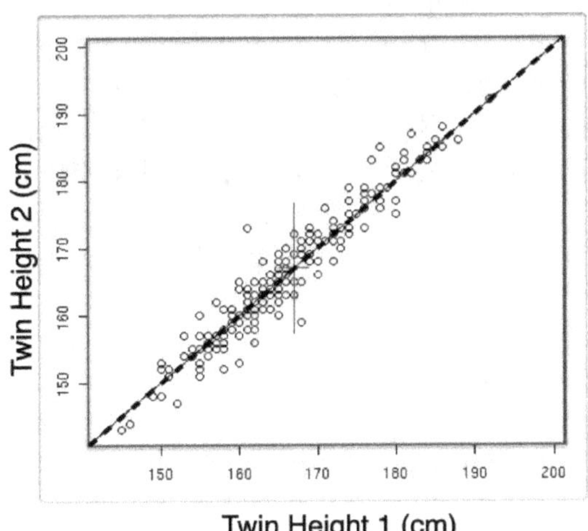

1.6.3. *Measurement of Quantum Entanglement and Correlation*

Let us now consider a quantum mechanical system, where a pair of quantum particles are generated in an entangled state. If what is entangled is the particles' direction of oscillation along the axes x and y, then the quantum state of the combined system can be written as shown below:

$$|\Psi\rangle = \frac{1}{\sqrt{2}}(|1_x\rangle|2_x\rangle + |1_y\rangle|2_y\rangle).$$

This is one example of an entangled state on this basis. Another example could be $|\Psi\rangle = 1/\sqrt{2}(|1_x\rangle|2_y\rangle + |1_y\rangle|2_x\rangle)$.

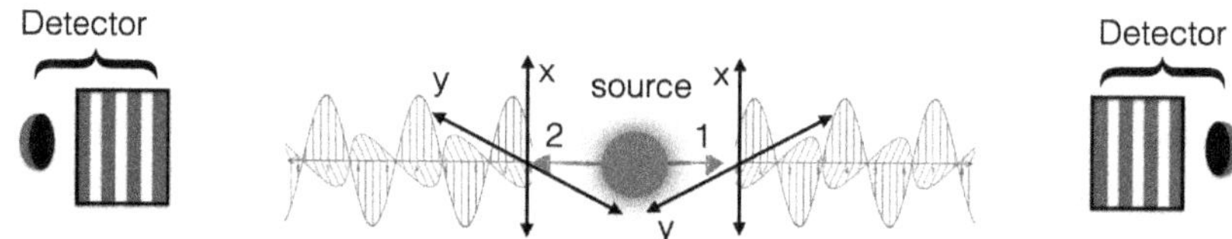

To verify that the two particles are entangled, we need to generate copies of the particles, perform a series of measurements and repeat the process until we have enough data to verify the entanglement in statistical analysis. In terms of measurements, we have had similar examples before, where we introduced detectors that can be set to measure oscillations along particular directions. If a detector on the right-hand side measures the state of particle 1 which is aligned along the x axis, it will only click when particle 1 oscillates along the x direction. In this case, assuming the state of the system is properly described by Ψ above, the two detectors will simultaneously click if both are set to measure particles in the x or y basis. In all real-world experiments, the state of the particles is never pure, and the system is always in a mixture of different states due to noise or loss. As a result, the degree of entanglement can vary depending on experimental conditions. By repeating the measurements over many similarly produced particles, the probability of coincident detection in different bases can be obtained, from which the degree of entanglement can be deduced.

1.6.4. *Bell Inequality*

Bell in 1964 came up with a measure of entanglement shown as S parameter below which is calculated from the probability distribution of different events.

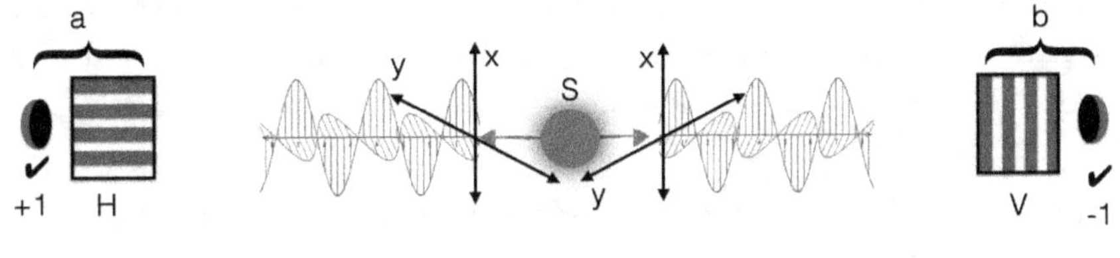

$$S = E(a, b) - E(a, b') + E(a', b) + E(a', b')$$

$$E(a, b) = P_{++}(a, b) + P_{--}(a, b) - P_{+-}(a, b) - P_{-+}(a, b)$$

Bell showed that for any classical system, $-2 \le S \le 2$. Violating this inequality proves non-classical correlation.

The quantity S above is defined based on probabilities of detection in different bases. Consider the same example as before where oscillation directions of the two particles are entangled. The two detectors on the left and right can be set along the horizontal or vertical directions, or directions diagonal to H and V. The letters a & b and a' & b' refer to left–right detectors set to measure particles in H/V and diagonal basis, respectively. For example, when the left and right detectors are set to detect particles oscillating in the horizontal direction, the mean number of coincident clicks normalized to total clicks provides a measure of conditional probability $P_{++}(a, b)$. Similarly, when the left and right detectors are set to detect particles oscillating in the same diagonal directions, the mean number of normalized coincident clicks provides a measure of conditional probability $P_{++}(a', b')$ or $P_{--}(a', b')$. We can see that 16 sets of measurements on different basis are needed to quantify S. Bell's theorem states that S for any classical correlation has to be bounded by ± 2. Violating this inequality indicates the presence of entanglement (higher-than-classical correlations) between two or more particles. The experimental setup above can be used to verify entanglement between photons entangled on a polarization basis. For two entangled atoms in the energy basis, for example, the state of each atom in each of the energy states as well as the superposition of energy states needs to be measured to quantify the degree of entanglement. Most entangling sources are not perfect and the degree of entanglement can vary from system to system due to environmental noise, loss, cross-talk, and multi-particle events. The quantification of entanglement using Bell inequality is essential to verify the degree of entanglement in many systems.

1.6.5. *Generation of Quantum Optical Entanglement*

We can provide an example to see how entangled particles can be generated. Specifically, let us describe the generation process of entangled photon pairs using nonlinear crystals.

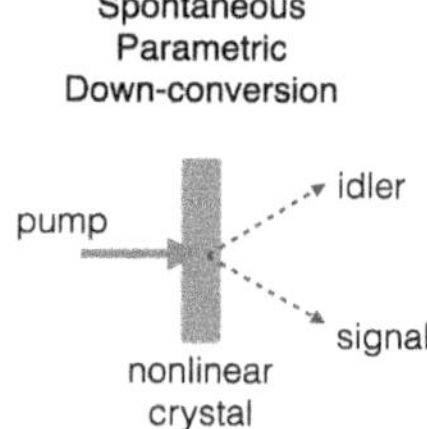

Electric polarization inside a nonlinear crystal is given by:

$$P = \epsilon_0 \chi^{(1)} E + \epsilon_0 \chi^{(2)} E.E + \epsilon_0 \chi^{(3)} E.E.E + \cdots$$

second-order   third-order

nonlinearity   nonlinearity

(e.g. parametric   (e.g. four-wave

downconversion)   mixing)

The electric polarization in materials created as the result of interaction with an electromagnetic field or light can linearly or nonlinearly depend on the amplitude of the E-field. When the nonlinear coefficients of a material, $\chi^{(2)}$ and $\chi^{(3)}$ are not zero, we deal with an optical nonlinear material. Such nonlinearity can be the result of parametric down-conversion (in the case of non-zero $\chi^{(2)}$) or four-wave mixing processes (in the case of non-zero $\chi^{(3)}$) in the material. In the down-conversion process, for example, an incident pump photon undergoes scattering generating two photons (signal and idler photons) while conserving total energy and momentum. The polarization, P, in the material that gives rise to the down-converted photons nonlinearly depends on E-field amplitudes. In other words, by replacing $E = E_\text{s} + E_\text{i}$, the equation above will have terms that are proportional to $E_\text{s} \cdot E_\text{i}$ meaning polarization of signal (E_s) and idler (E_i) fields cannot be described separately or independently. This results in correlations between idler and signal photons that are both generated from the effective oscillating electric dipole, P. When the classical nature of these fields is considered, the process can be described by the Maxwell equations. In the quantum regime, where the quantized nature of particles (photons) are also considered, correlations in photon number (particle property), as well as correlation in field fluctuation (wave property), can be observed. The size of fluctuation in each signal and idler path is limited by the Heisenberg Uncertainty Principle. However, when observing correlations, the relative fluctuation between the two paths can go below the Heisenberg uncertainty limit. In other words, the correlation can surpass the maximum possible classical correlations.

Through a down-conversion process, a single photon of angular frequency ω_0 simultaneously generates a pair of signal and idler photons of angular frequencies ω_1 and ω_2.

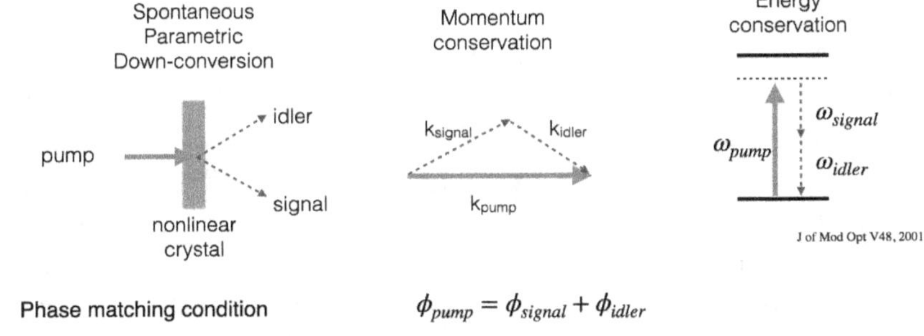

For a nonlinear material (e.g. a barium-borate or BBO crystal) to generate entangled pairs, some conditions need to be satisfied. First, the material needs to be transparent at the frequencies of the pump, idler, and signal. Also, the strength of the nonlinearity needs to be large enough to suppress the classical noise associated with background photons. Importantly, the phase matching condition must also be satisfied. The phase of any electromagnetic field is defined by its frequency and wave vector, i.e. $\phi = \omega t + \mathbf{k}.\mathbf{r}$, where $\mathbf{r}$ is the propagation length of the field with wave vector, $\mathbf{k}$. For the total phase of the input light to be equal to the total phase of the output light (continuity principle), both energy and momentum conservation conditions need to be met.

As seen above, the momentum conservation indicates that the total momentum of the particles (pump, signal, and idler photons) must be conserved. This condition limits the angles through which the signal and the idler are emitted. The energy conservation indicates that the frequency of the pump photon must be equal to the sum of the output photons' frequencies. This does not necessarily mean that the frequencies of the signal and idler photons should be equal to half of the pump frequency. It is possible to design experiments to achieve a degenerate (or non-degenerate) down-conversion process where the idler/signal frequencies are equal (or not equal). Later in this chapter, we will discuss other aspects of the entangled pair generation process, such as noise and probability.

1.6.6. *Generation of Matter–Matter Entanglement*

Photon–photon entanglement is routinely generated using nonlinear materials and is used to perform quantum sensing/communication/ simulation tasks. Moreover, matter–matter entanglement is of great importance for quantum communication and computation. Such entanglement can be achieved by tailoring nonlinear interactions between atoms, for example, through phonon/photon/electron-mediated interactions. Two or more qubits with the ground, $|g\rangle$, and excited, $|e\rangle$, energy states can interact directly (e.g. via dipole–dipole interaction) or indirectly (e.g. via electromagnetic radiation) to create entanglement between the qubits. The strength of nonlinear interactions between quantum particles determines the probability of creating entanglement.

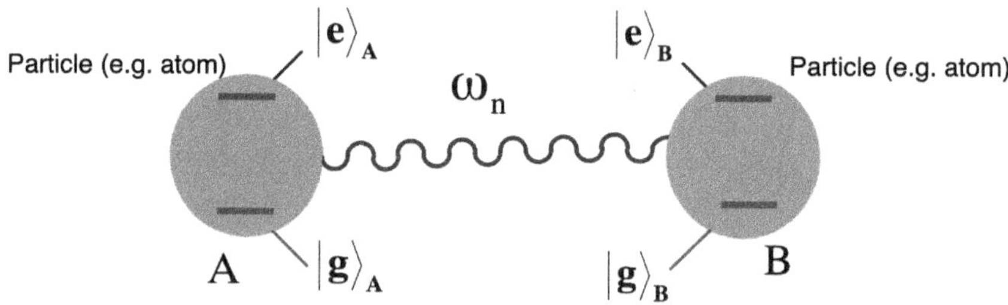

The nonlinearity of the entangling process can also be introduced through the measurement process itself. For example, two non-interacting atoms can be entangled by performing measurements on emitted photons from atoms. If the emission probability is small, each atom remains dark most of the time but occasionally emits a photon. For an atom initially ($t = 0$) prepared in the excited or a meta-stable state, the state of the atom–photon at $t > 0$ can then be written as $A|g\rangle|1\rangle + B|e\rangle|0\rangle$. If the photon is emitted (i.e. photon state $|1\rangle$) atom decays to the $|g\rangle$ and if no photon is emitted, (i.e. photon state $|0\rangle$), the atom remains in the excited state, $|e\rangle$. The emission process can be triggered by another laser via, for example, the Raman process, with a limited transition probability. It can be shown that by interfering the emission signals of the two atoms followed by coincident detection, the two atoms can be probabilistically entangled. We will discuss similar entangling processes in more detail in the following chapters.

Chapter 2

Photons and Atoms for Quantum Optical Information

2.1. Quantum Electromagnetic Fields

In this chapter, we introduce two of the most important resources used to build quantum networks: photons and atoms. Understanding classical and quantum descriptions of electromagnetic (EM) fields is the first step in unraveling their dynamics and interactions with matter. A photon is a packet of EM field with $h\nu$ of energy, where ν is the EM field oscillation frequency and h (or $2\pi\hbar$) is the Planck constant. A laser or other sources of EM fields can generate pulses of light containing many photons. The propagation and interaction of such classical fields can be described by the well-known Maxwell equations. Attenuating such fields to the single-photon level produces quantum particles whose full description requires the quantum theory of the EM field. Optical photons, generated simply by attenuating laser pulses or by using nonlinear interactions of intense light pulses with atoms and crystals, can carry quantum information. In this section, we provide an overview of quantum EM fields, their generation, and their properties.

2.1.1. *Classical EM Theory — Maxwell's Equations*

Maxwell's equations summarize the combined electric and magnetic response of a medium:

$$\nabla \cdot \boldsymbol{D} = \varrho, \quad \text{Gauss's law for electrostatics}$$

$$\nabla \cdot \boldsymbol{B} = 0, \quad \text{Equivalent of Gauss's law for magnetostatics}$$

$$\nabla \times \boldsymbol{\mathcal{E}} = -\frac{\partial \boldsymbol{B}}{\partial t}, \qquad \text{Faraday–Lenz Law}$$

$$\nabla \times \boldsymbol{H} = \boldsymbol{j} + \frac{\partial \boldsymbol{D}}{\partial t}, \qquad \text{Amperes's law}$$

The equations provide the mathematical model for today's technologies based on electric, optical, and radio signals.

The classical theory of electromagnetics describes the propagation and interaction of classical electromagnetic fields in great detail. When only the wave nature of the fields is considered, the Maxwell equations accurately capture the dynamics of electric and magnetic fields. Take, for example, the Faraday–Lenz Law and Ampere's Law provided above. The symbols E and B represent electric and magnetic field vectors, respectively. Also, permittivity ϵ_0, speed of light c, and current density j are defined for fields in a vacuum. Electromagnetic fields travel at the speed of light and the time variation in one causes a curly spatial variation in the other one and vice versa.

2.1.2. *Classical EM Theory — Solution to Maxwell's Equations*

The typical solutions to Maxwell's equations in 1D are transverse waves with E and B fields at right angles:

$$E_x(z,t) = E_0 e^{-i\omega t + ikz + \phi} = f(t) \cdot u(z)$$

$$B_y(z,t) = B_0 e^{-i\omega t + ikz + \phi} = h(t) \cdot v(z)$$

where the wave vector

$$k = \frac{2\pi}{\lambda} = \frac{\omega}{v} = \frac{n\omega}{c}.$$

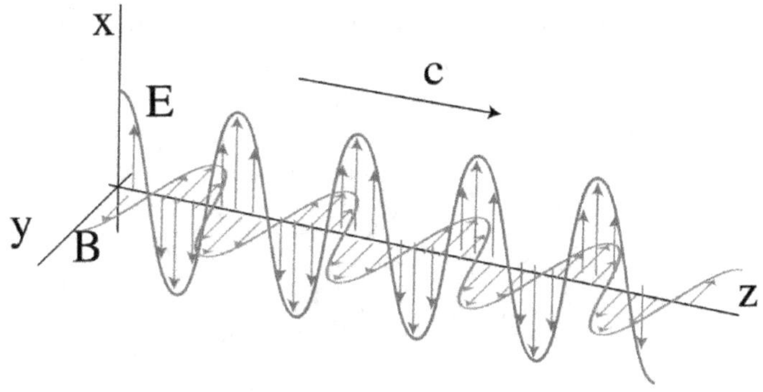

A typical solution to the Maxwell equations above is for a unidirectional field (plane waves), E_x and B_y, where $k = 2\pi/\lambda$ is the wavenumber and ω, λ and ϕ are the angular frequency, wavelength, and phase of the field propagating in the z direction. A more general solution can be expressed in terms of a temporal mode (describing time dependency), $f(t)$, and a spatial mode (describing spatial dependency), $u(z)$, of the fields.

2.1.3. *Quantum EM Theory*

The classical Hamiltonian or energy of an electromagnetic field (plane-wave approximation) in space can be expressed in terms of the volume integral of the fields.

$$E_x(z,t) = E_0 e^{-i\omega t + ikz + \phi} = f(t) \cdot u(z)$$

$$B_y(z,t) = B_0 e^{-i\omega t + ikz + \phi} = h(t) \cdot v(z)$$

$$H = \frac{1}{8\pi} \int (E^2 + B^2) dV = \frac{1}{8\pi}[f(t)^2 + k^2 h(t)^2].$$

Now, compare the field Hamiltonian above with the Hamiltonian of a harmonic oscillator, $H_{H.O.}$, where $p(t)$ and x are momentum and position and m and ω are the mass and the angular frequency of an oscillating object or particle. It can be seen that the two Hamiltonians have similar forms. So, the mean energy of the EM field should resemble that of a harmonic oscillator.

$$H_{H.O.} = \frac{1}{2}\left(\frac{P(t)^2}{2m} + m\omega^2 x^2\right)$$

$$\langle H \rangle_{H.O.} = E_n = \left(n + \frac{1}{2}\right)\hbar\omega$$

To arrive at the quantum form of these Hamiltonians, one needs to replace the dynamical variables in classical theory, e.g. $p(t)$ and x, with their corresponding quantum mechanical operators, e.g. $\hat{p} = -i\hbar\frac{d}{dx}$ and $\hat{x}$. Here, $\hbar$ is the reduced Planck constant. In the case of a Harmonic oscillator, the eigenvalue equation is then written as

$$\hat{H}|\psi\rangle = E|\psi\rangle$$

where $|\psi\rangle$ is the eigenstate or wave function of the particle and E is the expectation value of its energy. We can substitute $\hat{p} = -i\hbar\frac{d}{dx}$ and $\hat{x}$ in the equation for $H_{H.O.}$ and obtain the Hamiltonian operator. We can show that there exist multiple solutions to the eigenvalue equation above in the forms of Hermit polynomial functions distinguished by integer numbers n. The mean energy of the particle in a Harmonic potential is then given by $\langle H \rangle = E = (n + \frac{1}{2})\hbar\omega$, where we refer to the integer n as the quantized number of excitations (phonons) that the particle has.

2.1.4. *Quantized EM Fields*

State of the field with n quanta of energy is represented by $|n\rangle$ which are the eigenfunctions of the Hamiltonian.

$$\hat{H} = \left(\hat{a}^\dagger\hat{a} + \frac{1}{2}\right)\hbar\omega$$

$$\langle\hat{a}^\dagger\hat{a}\rangle = \langle\hat{n}\rangle = n.$$

Equivalently, for the EM field, we can define

$$\hat{x} = \sqrt{\frac{\hbar}{2m\omega}}(\hat{a} + \hat{a}^\dagger), \quad \hat{p} = -i\sqrt{\frac{\hbar m\omega}{2}}(\hat{a} - \hat{a}^\dagger), \quad \hat{n} = \hat{a}^\dagger\hat{a},$$

$$\hat{H} = \left(\hat{n} + \frac{1}{2}\right)\hbar\omega$$

where $\hat{n}$ is the number operator and $\hat{a}$ and $\hat{a}^\dagger$ are ladder operators which decrease or increase the excitation of the system by one quanta of energy when applied to the original state or wave function.

By replacing dynamical variables $f(t)$ and $h(t)$ with operators we can arrive at quantum representation of E (and similarly for B) field:

$$E(z,t) = i\sqrt{\frac{\hbar\omega}{2V\epsilon_0}}[\hat{a}(t)e^{ikz} - \hat{a}^\dagger(t)e^{-ikz}]u(x)$$

When considering an electromagnetic field confined in a region of space with volume V and permittivity ϵ_0, the photons behave like particles in a harmonic trap. This is one way to understand the resemblance of the two Hamiltonians H_{HO} and H_{EM}. Such wave–particle duality nature of quantum particles (or waves) is at the core of quantum theory differentiating quantum descriptions from classical ones.

The energy of an electromagnetic field $\hat{H} = (\hat{n} + \frac{1}{2})\hbar\omega$ results from the n quanta of field excitations (or photons). The electric field operator corresponding to a single-photon wave packet within volume V of space can also be written in terms of ladder operators $\hat{a}$ and $\hat{a}^\dagger$, which respectively correspond to the lowering or raising of the field energy by one quanta, as

$$\hat{E}_x = \sqrt{\frac{\hbar\omega}{2V\epsilon_0}}[\hat{a}^\dagger e^{-i\omega t} + \hat{a}e^{i\omega t}]u(z)$$

where $u(z)$ is a function describing the spatial envelope of the field. Note that this is the operator of a single-mode field, i.e.

a monochromatic field traveling along a single direction, z (plane wave). For a multi-mode field, one needs to sum over each frequency (spectral modes), directions (spatial modes) and time window (temporal modes) that a photon lives in.

The electric and magnetic field operators for multi-spatial mode EM field can be written as

$$\hat{\mathbf{E}} = \sqrt{\frac{\hbar}{2\epsilon_0 V}} \sum_k i\epsilon_k \sqrt{\omega_k}\,\hat{a}_k u_k(r) e^{-i\omega_k t} + h.c.$$

$$\hat{\mathbf{H}} = \sqrt{\frac{\hbar}{2\epsilon_0 V}} \frac{1}{c\mu_0} \sum_k i(\hat{k} \times \epsilon_k)\sqrt{\omega_k}\,\hat{a}_k u_k(r) e^{-i\omega_k t} + h.c.$$

where $u_k(r) = \frac{e^{-i\mathbf{k}.\mathbf{r}}}{\sqrt{2\pi}}$ is the spatial mode, $\omega_k = c|k|$, ϵ_k is a unit polarization vector, and $h.c.$ indicates the Hermitian conjugate terms corresponding to the first term.

2.1.5. *Field Quantization*

To show field quantization we previously used the analogy with the harmonic oscillator to derive the quantization of energy and ladder operators. Here we derive field quantization from the Maxwell equations.

Consider the magnetic field, B written in terms of the vector potential, A, where

$$B = \nabla A.$$

Here the electric field E can be written in terms of a scalar potential ϕ and vector potential A

$$E = -\nabla\phi - \frac{1}{c}\frac{\partial A}{\partial t}$$

where c is the speed of light in the vacuum. Using the Maxwell equation and for zero current density and charge we can write the following wave equation

$$\nabla^2 A - \frac{1}{c^2}\frac{\partial^2 A}{\partial t^2} = 0.$$

The solution to this wave equation for a field confined to volume V is given by

$$A(r, t) = \frac{1}{\sqrt{V}} A_0 e^{-i\omega t} e^{i\mathbf{k}\cdot\mathbf{r}}$$

where ω is the frequency of oscillation and $\mathbf{k}$ is the wavevector. We define the temporal mode as $A(t) = A_0 e^{-i\omega t}$. To obtain the real part of the solution, $Re[A(r,t)] \equiv A(r,t)$, we write

$$A(r, t) = \frac{1}{2\sqrt{V}}(A(t)e^{i\mathbf{k}\cdot\mathbf{r}} - A^*(t)e^{-i\mathbf{k}\cdot\mathbf{r}}).$$

We can then write the electric field as

$$E(r, t) = \frac{i}{2c\sqrt{V}}(\omega A(t)e^{i\mathbf{k}\cdot\mathbf{r}} - \omega A^*(t)e^{-i\mathbf{k}\cdot\mathbf{r}}).$$

In the case of multimode field (see the figure below), this equation changes to

$$E(r, t) = \frac{i}{2c\sqrt{V}} \sum_k (\omega_k A_k(t)e^{i\mathbf{k}\cdot\mathbf{r}} - \omega_k A_k^*(t)e^{-i\mathbf{k}\cdot\mathbf{r}}).$$

The field energy or classical Hamiltonian can be written as

$$H = \frac{1}{8\pi} k^2 |A_k|^2.$$

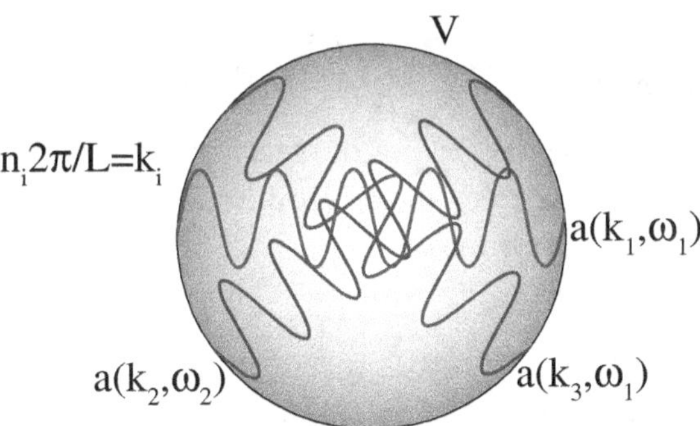

EM field quantization in mode volume, V.

We now define the quantum operators to replace the dynamical variables $A_k(t)$ and $A_k^*(t)$ as

$$A_k(t) \;\to\; 2c\sqrt{\frac{\hbar}{2\omega_k \epsilon_0}}\,\hat{a}_k(t)$$

$$A_k^*(t) \;\to\; 2c\sqrt{\frac{\hbar}{2\omega_k \epsilon_0}}\,\hat{a}_k^\dagger(t).$$

The factor $2c\sqrt{\frac{\hbar}{2\omega_k \epsilon_0}}$ is considered to simplify the Hamiltonian. Here $\hbar$ is the Plank constant and ϵ_0 is the vacuum permittivity.

We can now write the electric field in the quantum form

$$\hat{E}(r,t) = i\sqrt{\frac{\hbar}{2\epsilon_0 V}} \sum_k (\sqrt{\omega_k}\,\hat{a}_k(t)e^{i\mathbf{k}.\mathbf{r}} - \sqrt{\omega_k}\,\hat{a}_k^\dagger(t)e^{-i\mathbf{k}.\mathbf{r}}).$$

This is similar results when compared to the method using the harmonic oscillator analogy.

2.1.6. *Quantized Field Operators*

The dynamical variables E and B in the classical EM theory can be replaced by the corresponding ladder operators in quantum theory to write the quantum Hamiltonian and the equations of motion.

Some of the definitions and properties related to the ladder operators are summarized below.

- Number operator $\hat{n}$ $\hat{n} = \hat{a}^\dagger \hat{a}$
- Creation operator $\hat{a}^\dagger$ $\hat{a}^\dagger|n\rangle = (n+1)^{1/2}|n+1\rangle$
- Annihilation operator $\hat{a}$ $\hat{a}|n\rangle = (n)^{1/2}|n-1\rangle$

- Photons are quantized packets of electromagnetic waves.
- Photons can be created or annihilated upon interaction with external stimuli/environment.
- The order of operation matters.

The photon number operator (corresponding to the dynamical variable photon number) can be expressed in terms of the ladder operators. We note that the photon number is proportional to the classical intensity of the EM field or its energy. In the quantum formalism, the quantum Hamiltonian of the EM field can be written in terms of the number operator, $\hat{a}^\dagger\hat{a}$.

The ladder operators do not commute: $[\hat{a}, \hat{a}^\dagger] = 1$. This suggests that the order of the operations is important.

2.1.7. *Coherent States of Light*

When measuring different properties of EM fields (e.g. amplitude and phase), one can see fluctuations around the mean value, when minimum, equal to the vacuum fluctuations. The fluctuations can be seen in the form of amplitude, phase, or intensity fluctuations and are caused by intrinsically random processes involved in the generation and interaction of EM fields. The randomness in such processes is theoretically captured by the Heisenberg uncertainty principle for non-commuting variables, such as amplitude and phase. States of EM fields with the smallest possible amount of classical noise are called minimum uncertainty states or coherent states. Coherent states lie at the boundary between classical and quantum mechanical regimes. A laser beam is a good approximation of a coherent state that can be easily produced for the semi-classical study of quantum optics.

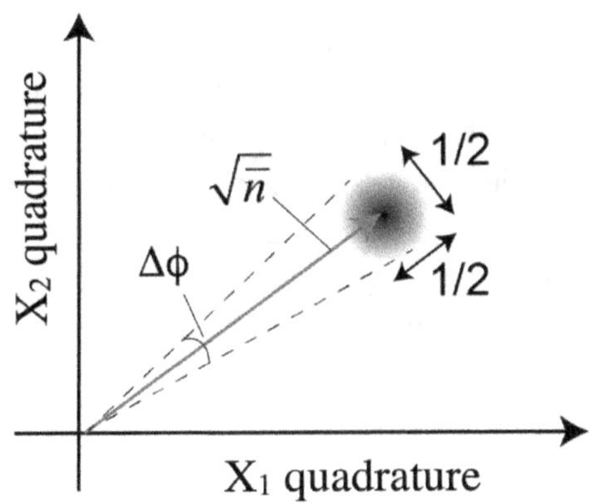

Instead of photon number, we can consider amplitude and phase fluctuations of the field. To theoretically model the fluctuations and their dynamics, we define field "quadratures". The amplitude and phase quadratures of EM fields can be defined in terms of ladder operators.

We can define X_1 and X_2 quadratures of the field that correspond to the sine and cosine parts of the time-dependent electric field:

$$\text{Amplitude quadrature: } X_1(t) = \frac{\hat{a} + \hat{a}^\dagger}{2} \propto E \cdot \cos(\omega t + \phi(t))$$

$$\text{Phase quadrature: } X_2(t) = \frac{\hat{a} - \hat{a}^\dagger}{2i} \propto E \cdot \sin(\omega t + \phi(t))$$

$$\text{Uncertainty: } \Delta X_2 \Delta X_1 \geq \frac{1}{4}$$

These quadratures satisfy the commutation relation

$$[X_1, X_2] = i.$$

Since the two operators $\hat{X}_1$ and $\hat{X}_2$ do not commute, there exists a corresponding Heisenberg relation written as $\Delta X_1 \Delta X_2 \geq 1/4$, where $\Delta X_{1,2} = \sqrt{\langle \hat{X}_{1,2}^2 \rangle - \langle \hat{X}_{1,2} \rangle^2}$. The Heisenberg relation indicates that certainty in the knowledge of one quadrature corresponds to uncertainty in the knowledge of the other. Although the uncertainty principle is valid for both quantum and classical signals, the size of the fluctuations compared to the magnitude of a typical classical EM field is negligible. In the quantum case, we typically deal with single-photon-level fields and as the minimum intensity fluctuation (also shot noise) scales inversely with the square root of the mean photon number, the size of the fluctuation is comparable to the mean signal amplitude in the single-photon-level light and cannot be ignored.

For coherent states, we have $\Delta X_1 = \Delta X_2 = 1/2$, i.e. the amplitude and phase fluctuations have the same standard deviation.

The figure above shows the phase-space representation of a coherent state (ball-on-stick representation) with a mean photon number $\bar{n}$. The phase ϕ is referenced to some reference field or a local oscillator field.

2.1.8. *Energy–Time Uncertainty*

When the energy of an EM field is measured instead of its quadratures (amplitude and phase), an equivalent uncertainty principle can be derived from X_1–X_2 relation as

$$\Delta E \Delta t \geq \hbar/2$$

where Δt represents the uncertainty in the arrival of the EM field. By replacing $\Delta E = \Delta n \hbar \omega$ and $\Delta t = \Delta \phi / \omega$, we can equivalently write

$$\Delta n \Delta \phi \geq 1/2.$$

This uncertainty relation suggests that a single photon with known n does not have a well-defined phase. We note that X_1 and X_2 describe the continuous or wave-like property of an EM fields. The expectation value of these operators can be measured by recording the phase and amplitude of the EM field, which are continuous variables (can take any real value and are not quantized).

To quantify the energy and time uncertainty relationship, we can detect the energy (or frequency) and arrival time of the photon. The energy–time uncertainty principle is also shown above. Although the energy of a photon is a discrete variable and is quantized, its value is not well determined, unless the photon is delocalized in time.

2.1.9. *Homodyne Measurement*

The phase and amplitude of EM fields can be measured with high precision using the homodyne detection scheme shown below where a signal ($\mathcal{E}$) is interfered with a reference EM field or a local oscillator (E_{LO}).

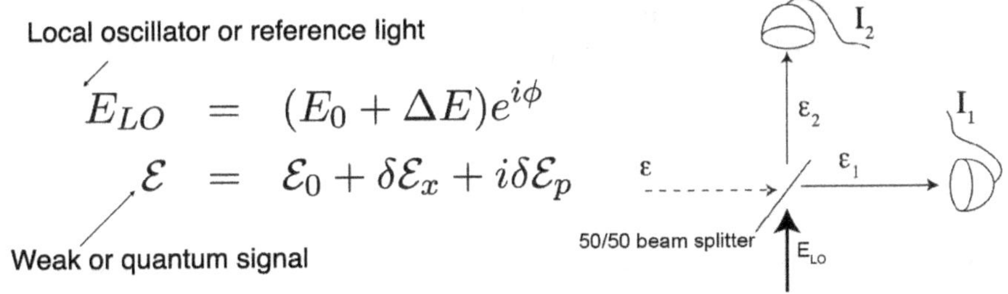

It can be shown that the subtracted intensity of two photodetectors is given by:

$$I_2 - I_1 \simeq 2E_0\delta\mathcal{E}_x\cos(\phi) + iE_0\delta\mathcal{E}_p\sin(\phi).$$

In the above definitions for $\mathcal{E}$ and E_{LO}, E_0 and $\mathcal{E}_0$ are the mean amplitudes of the LO and signal light, ΔE is the local oscillator (LO) amplitude fluctuation, $\delta\mathcal{E}_x$ and $\delta\mathcal{E}_p$ are the amplitude and the phase fluctuations of the signal field, and ϕ is the phase difference between the two fields.

Assuming the local oscillator is much stronger than the signal, the subtracted current of the two photodetectors is the homodyne signal that is proportional to the weak signal fluctuations. This can be shown by writing $\mathcal{E}_1 = 1/2(E_{\mathrm{LO}} - \mathcal{E})$ and $\mathcal{E}_2 = 1/2(E_{\mathrm{LO}} + \mathcal{E})$ to calculate $I_2 - I_1 = |\mathcal{E}_2|^2 - \mathcal{E}_1|^2$.

When $\phi = 0$ or $\phi = \pi/2$, the current difference is proportional to amplitude or phase fluctuations, respectively. We see that the current difference is insensitive to the intensity fluctuations of the LO and it measures the signal fluctuation, which is amplified by a factor of the square root of the LO intensity. Homodyne measurement enables the measurement of signal fluctuations down to single-photon precision.

2.1.10.   *Coherent State in Number Basis*

Photon number statistics of optical states can also be determined by performing single-photon detection.

Number basis are eigenstates ($|n\rangle$, state with n quanta of energy) of the Hamiltonian:

$$\hat{H} = \left(\hat{a}^\dagger\hat{a} + \frac{1}{2}\right)\hbar\omega$$

A coherent state (represented in Dirac notation as $|\alpha\rangle$) can be written as a superposition of a number of states (represented in Dirac notation as $|n\rangle$) as

$$|\alpha\rangle = e^{-|\alpha|^2/2}\sum_{n=0}^{\infty}\frac{\alpha^n}{\sqrt{n!}}|n\rangle$$

where $n = |\alpha|^2$ is the mean photon number and $|n\rangle$ is an optical state with exactly n photons, known as Fock state. For a vacuum we have $|n = 0\rangle$ that is a coherent state with $\alpha = 0$ Defining the mean photon number a coherent state as $|\alpha|^2 = \bar{n}$ The photon number fluctuation or variance of the photon number can be written as $(\Delta n)^2 = (\bar{n} + \bar{n}^2) - \bar{n}^2 = \bar{n}$. Here $\bar{n}$ is the same as classical n defined earlier in this chapter.

Single-photon detectors produce clicks when a photon hits the detector (with certain efficiency). As long as the two photons do not arrive at the same time or more precisely within the detector "dead" time, the mean number of photons within a detection time window can be determined by summing over all the clicks in that time window. As the photon number of a coherent pulse is not well determined, every time a laser fires a pulse, it will have a different number of photons. The distribution of photon numbers in a coherent state is a random distribution given by the Poisson distribution.

A characteristic of such distribution is that the photon number fluctuation is equal to the square root of the mean photon number.

2.1.11. *Beam Splitter in Quantum Optics*

In this section, we mathematically describe the beam splitter operation in the quantum regime where single photons can interfere as particles and waves. A beam splitter is a linear optical component, where its output modes are a linear combination of the input modes. We consider a two-port optical beam splitter shown in the following figure with reflection and transmission coefficients r and t, respectively. Beam splitters are asymmetric in the sense that the reflection of the two input modes a and b are not the same due to the refractive index contrast seen by each mode at different input ports. The phase of one mode changes by $\pi/2$ and the other one by $-\pi/2$ causing a phase difference of π between the reflection modes at the output.

The classical output fields can be written in terms of r and t, where $|r|^2 + |t|^2 = 1$ assuming a loss-less beam splitter, as

$$b_{out} = r a_{in} + t b_{in}$$

$$a_{out} = -r b_{in} + t a_{in}$$

where intensity of output modes can be calculated as $I_a = |a_{out}|^2$ or $I_b = |b_{out}|^2$.

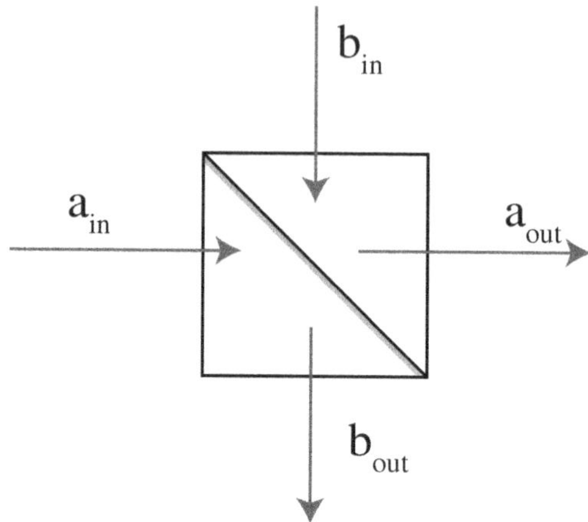

Schematic of a beam splitter with two input and two output ports.

In the quantum regime, we replace the field amplitudes with operators

$$\hat{b}_{out} = r\hat{a}_{in} + t\hat{b}_{in}$$

$$\hat{a}_{out} = -r\hat{b}_{in} + t\hat{a}_{in}$$

$$\hat{b}^{\dagger}_{out} = r\hat{a}^{\dagger}_{in} + t\hat{b}^{\dagger}_{in}$$

$$\hat{a}^{\dagger}_{out} = -r\hat{b}^{\dagger}_{in} + t\hat{a}^{\dagger}_{in}.$$

We can write the beam splitter operation in a matrix form, given the 2D space of a beam splitter

$$\begin{pmatrix} \hat{b}_{out} \\ \hat{a}_{out} \end{pmatrix} = \begin{pmatrix} r & t \\ t & -r \end{pmatrix} \begin{pmatrix} \hat{a}_{in} \\ \hat{b}_{in} \end{pmatrix}.$$

Rearranging terms we can find input modes in terms of the output modes

$$\hat{b}_{in} = -r\hat{a}_{out} + t\hat{b}_{out}$$

$$\hat{a}_{in} = r\hat{a}_{out} + t\hat{b}_{out}$$

$$\hat{b}^{\dagger}_{in} = -r\hat{a}^{\dagger}_{out} + t\hat{b}^{\dagger}_{out}$$

$$\hat{a}^{\dagger}_{in} = t\hat{a}^{\dagger}_{out} + r\hat{b}^{\dagger}_{out}.$$

Let us consider a single photon at the input mode a and vacuum input mode b. The input state of the two ports can be written as

$$|\psi_{in}\rangle = |1_a\rangle|0_b\rangle = \hat{a}^\dagger|0_a\rangle|0_b\rangle.$$

Replacing $\hat{a}^\dagger$ in the above equation with $t\hat{a}^\dagger + r\hat{b}^\dagger$, we can find the output state of the beam splitter as

$$|\psi_{out}\rangle = (t\hat{a}^\dagger + r\hat{b}^\dagger)|0_a\rangle|0_b\rangle = t|1_a\rangle|0_b\rangle + r|0_a\rangle|1_b\rangle.$$

The probability of detecting a photon in output port b, for example, becomes

$$P_b = \langle\psi_{out}|\hat{b}^\dagger_{out}\hat{b}_{out}|\psi_{out}\rangle = |r|^2$$

which for a 50/50 beam splitter, i.e. $r = 1/\sqrt{2}$, the probability is 50%.

2.1.12. *Probability Distribution of Photons in Coherent States*

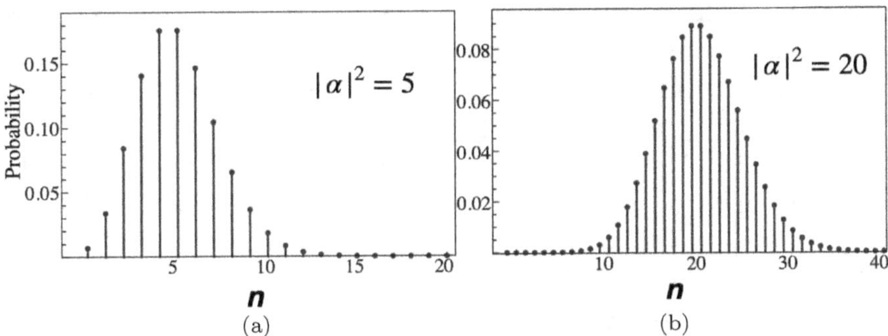

Probability of having n photons in the light beam is given by a Poisson distribution as

$$\boldsymbol{P}_n = |\langle n \,|\, \alpha\rangle|^2 = \mathrm{e}^{-|\alpha|^2}\frac{|\alpha^2|^n}{n!}$$

The expectation values and variance (square root of standard deviation) of the photon number in a Poisson distribution are equal. In comparison, the variance of the photon number in a thermal distribution is super-Poissonian following the Maxwell–Boltzmann distribution. The formula for the Poisson probability distribution is given above, where $|\alpha|^2$ is the shape parameter that indicates the average number of events in a given time interval and n is an integer.

The figure above is a plot of the Poisson probability density function for two values of $|\alpha|^2$ as a function of n for mean photon numbers of 5 and 20.

Squeezed states of EM fields exhibit fluctuations lower than vacuum noise. Squeezing is a unique feature of quantum states and can be in the form of quadrature, intensity, or photon-number squeezing. The squeezed states of light can be used for sensing beyond the classical limit, carrying information, or performing quantum information processing.

Heisenberg's uncertainty principle does not prohibit states from having sub-shot noise fluctuations. Nonlinear processes such as parametric amplification can produce sub-shot noise light fluctuation at the expense of increasing the fluctuations in the other non-commuting variables. For example, amplitude quadrature squeezing is accompanied by phase quadrature anti-squeezing and vice versa. These states are called squeezed states because the quantum noise in one quadrature becomes squeezed.

A photon-number squeezed state, $\Delta n < \sqrt{\bar{n}}$, has sub-Poisson photon-number statistics. In other words, its intensity noise is below the shot noise. Based on the Heisenberg uncertainty principle, the more certain the photon number or energy of the state, the less certain the phase of that state is.

2.1.13. *Quantum Optical Encoding*

Both continuous variable (CV) quantum optical states such as squeezed states and discrete variable (DV) states such as number states or Fock states can carry quantum information. In the case of CV states, correlations in continuous variable parameters such as amplitude and phase of EM fields are used to encode information. The non-classical correlation translates into sub-shot-noise fluctuations in amplitude and phase quadratures. In the case of DV states, discrete parameters such as photon number measured on various basis (e.g. polarization, frequency, time, etc) are used for encoding quantum information. The non-classical correlation in this case can be evaluated by measuring the cross-correlation function, that when taking values less than 1 or larger than 2 represent quantum signatures in the signal.

The simplest forms of quantum optical states are Fock states or number states that can be written as a superposition of a well-defined number of particles or energy wave packets. The number basis $|N\rangle$, $N = 1, 2, \ldots$, is a complete orthonormal basis and eigenstates of the number operator $\hat{N} = \hat{a}^\dagger \hat{a}$.

A pure single-photon state can be described by a single number state $|1\rangle$ with zero photon-fluctuation or variance in the photon number, i.e. the photon number is well-determined. Based on the uncertainty principle, the phase of a single photon is completely undetermined. We note that this refers to the absolute phase of the photon. The phase of a photon, however, can be defined with respect to another photon.

As defined previously, the coherent states of light can be written in terms of the number basis. To mathematically describe a coherent state of amplitude $\alpha = \bar{n}$, where $\bar{n}$ is the mean photon number, in terms of vacuum state or zero-photon state $|0\rangle$

$$|\alpha\rangle = e^{-|\alpha|^2/2} e^{\alpha a_0^\dagger} |0\rangle.$$

As can be seen, there is a probability associated with having a certain photon number in a coherent state that depends on α. These probabilities give rise to fluctuations. In the case of coherent states, the fluctuations in either amplitude/photo-number and phase resulting from vacuum fluctuations are equal and minimum as allowed by Heisenberg's Uncertainty Principle.

The fluctuations of non-commuting variables, e.g. amplitude and phase quadratures, of the EM fields do not need to be equal as long as the product satisfies the Heisenberg Principle. It is possible, via nonlinear optical processes, to reduce quantum noise in one quadrature of a field (for example the phase quadrature) at the expense of increasing it in the complementary observable (i.e. the amplitude quadrature). These states are called quantum-squeezed states of light. Such states have applications in quantum sensing, quantum communication, and quantum computation. For example, a phase-squeezed light can be used to sense the position of an object upon reflection from the object. This can be done with precision beyond the classical limits. By interfering two light beams squeezed in opposite quadratures (one phase and one amplitude quadratures) CV entangled states can be created where reduced fluctuations in both amplitude and phase quadratures can be used to perform precision measurements.

The entanglement condition in the CV regime can be written [Duan *et al.* (2000)] as

$$\mathrm{Var}[\hat{X}_1 + \hat{X}_2] + \mathrm{Var}[\hat{P}_1 - \hat{P}_2] < 2$$

where Var here is the normalized variance and canonical quadrature operators for two fields 1 and 2, obey the commutation relation $[\hat{X}_{1,2}, \hat{P}_{1,2}] = i$. For a vacuum state we write $\mathrm{Var}[\hat{X}_{vac}] = \mathrm{Var}[\hat{P}_{vac}] = 1/2$. In what follows we normalize the variance to the vacuum fluctuation, so a variance of 1 means fluctuation is limited by the fundamental vacuum noise.

The squeezing operator is given by

$$\hat{S}(r, \theta, t) = \exp\left(\frac{r(e^{-2i\theta}\hat{a}(t)^2 - e^{2i\theta}\hat{a}^\dagger(t)^2)}{2}\right)$$

where r is the squeezing parameter describing the strength of noise reduction and θ is the squeezing quadrature angle describing amplitude quadrate (when $\theta = 0$), phase quadrature (when $\theta = \pi/2$) or any combinations of the two ($1 < \theta < \pi/2$). The squeezing factor, r, is real and positive and can be directly related to the standard deviation of the squeezed quadrature in the frequency domain as

$$\Delta\hat{X}_\theta(\omega) = e^{-r(\omega)}.$$

When the mean photon number is very small, the squeezed state is referred to as the squeezed vacuum state that is mathematically obtained by applying the squeezing operator on the vacuum state.

$$|0, r, \theta, t\rangle = \hat{S}(r, \theta, t)|0\rangle.$$

It is possible to create "bright" squeezed states of light where the mean photon number $\alpha > 0$. In this case, we define a displacement operator, $\hat{D} = e^{\alpha\hat{a}^\dagger - \alpha^*\hat{a}}$, to mathematically displace the squeezed vacuum state by amount an α from the origin. The displaced state is given by

$$|\alpha, r, \theta, t\rangle = \hat{D}(\alpha)\hat{S}(r, \theta, t)|0\rangle.$$

Adding noise or loss to quantum optical states (CV or DV) can wash out the quantum signatures. In the case of DV states, the loss of an optical channel is less problematic when compared to CV states. Consider two entangled photons propagating in two lossy channels. The loss reduces the probability or rate of the photons reaching the end nodes, however, once they arrive at the nodes they maintain their quantum correlations. Channel noise (e.g. polarization noise) however can reduce the fidelity and thus quantum signatures of DV states. CV states, on the other hand, are more sensitive to loss. The loss experienced by a squeezed state manifests itself as noise because less squeezing means more vacuum fluctuations in the mode of interest (that has more noise than the squeezed quadrature).

Consider two EM fields with photon number correlation or squeezing. We can write the noise reduction factor for a squeezed state as:

$$\eta = \frac{(\delta(n_1 - n_2))^2}{\bar{n}_1 + \bar{n}_2} = \frac{(\delta n_1)^2 + (\delta n_2)^2 - 2(\delta n_1)(\delta n_2)}{\bar{n}_1 + \bar{n}_2}$$

where δ represents the standard deviation in photon number and $\bar{n}$ is the mean photon number. For two coherent states, the numerator is the variance of the light that is equal to the mean photon number and in this case the noise reduction factor is 1.

In presence of loss ζ, the detected noise reduction factor becomes:

$$\eta_{det} = \zeta\eta + 1 - \zeta.$$

For example, light beams with an initial $10\,\mathrm{dB}$ of squeezing ($-10\,\mathrm{dB}$ of noise reduction) can provide $\eta = 0.1$ but after experiencing only 50% loss, the noise reduction factor quickly decreases to $\eta > 0.5$. When loss is more than 50%, it is difficult to see the effect of squeezing even when the initial squeezing is large.

Using this argument, one can show that cloning of quantum optical states like squeezed states is not possible [Wootters and

Zurek (1982)]. The cloning operation will add a vacuum of noise to conjugate variables and therefore the fidelity between the initial and final states will deteriorate. Considering this noise penalty, the "no-cloning limit" [Grosshans and Grangier (2001)] can be calculated, above which the output of a quantum device or channel is the best possible copy. To clone quantum optical information, an eavesdropper would need (at the simplest form) to intercept the transmission channel, split the light carrying information in half, and then amplify each half by a factor of two. Since splitting the light, e.g. using a beam splitter, introduces 50% loss to each output and amplification will amplify both the signal and its noise, one can show that the eavesdropper can obtain a copy of the state that at best overlaps with the initial state to about 30% (or 1/3). More accurately, if the end nodes of a transmission line can measure fidelity of more than 68% (or 2/3), they can confirm the channel's transmission security.

The fidelity of a quantum channel for Gaussian states (CV states with Gaussian noise or quadrature distribution) can be calculated using [Jeong *et al.* (2007)]

$$\mathcal{F} = \frac{2e^{-\frac{2\delta_X^2}{V_X+V_X'} - \frac{2\delta_P^2}{V_P+V_P'}}}{\sqrt{(V_X V_P' + 1)(V_P V_X' + 1)} - \sqrt{(V_X' V_P - 1)(V_X' V_P' - 1)}}$$

where $\delta_{X/P}$ is the coherent amplitude in phase space along the amplitude (X) and phase (P) axes. $V_{X/P}$ are the amplitude (X) and phase (P) quadrature variances at the input of the channel and V's are output variances.

One can see the fidelity measure depends on the size of the state $\delta_{X/P}$ and also its type. For DV state the fidelity is independent of efficiency of the channel. To characterize security of a quantum channel a state-independent measure known as T-V diagram can be used. This method was originally introduced for quantum non-demolition measurements [Poizat *et al.* (1994)] and later adapted to quantum communication [Lam (1998); Bowen *et al.* (2003)]. In this method, the conditional variance in the two quadratures is plotted against the signal transfer coefficient. The variance in quadratures

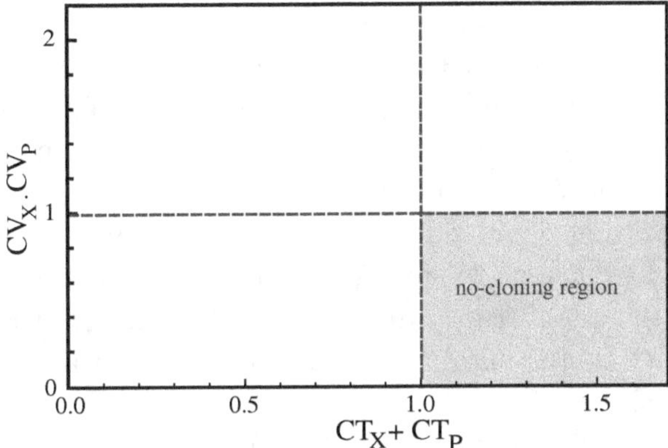

T-V diagram.

normalized to that of vacuum is a measure of the noise added by the channel or quantum memory to that quadrature.

The conditional transfer coefficient (CT) and conditional variances (CV) for two orthogonal quadratures are defined, respectively, as $CT_{X/P} = \eta/(1 + V'_{X/P} - V_{X/P})$ and $CV_{X/P} = (1 - T_{X/P})V'_{X/P}$. Here η is the channel transmission. In the limit of zero added noise, $CV_X = CV_P = 1$, for a coherent state. The amplitude and phase signal transfer coefficients $(CT_X$ and $CT_P)$ are a measure of channel or memory efficiency. One cannot measure amplitude and phase simultaneously with high precision. An eavesdropper copying the signal by dividing it in half then amplifying the signal by a factor of two (or reversed process) will introduce 100% more noise to the quadratures. So at best $CT_X = CT_P = 0.5$. Therefore, a quantum channel reaching $CV_X \times CV_P \leq 1$ and $CT_X + CT_P \geq 1$ reaches the quantum or "no-cloning limit" [Grosshans and Grangier (2001)] (see the figure above).

2.1.14. *Quasi-Probability Distribution*

Similar to the Maxwell-Boltzmann distribution of position and momentum of an ensemble of classical particles we can also define probability distribution functions for quantum particles representing the density matrix of the system of particles. Due to the non-commuting nature of quantum observables such as position and momentum or phase and amplitude quadrature of EM fields, these quantum probabilities can take negative values and thus we refer to them as quasi-probability distributions. One such quasi-probability functions is the Wigner function [Wigner (1932)] that can be experimentally constructed using measured quadrature values. In the CV regime, the quadrature value can be obtained from the homodyne measurements at a given phase difference between the local oscillator and the signal.

The Wigner distribution $W(X, P)$ for conjugate quadratures X and P, e.g. position and momentum, is defined as:

$$W(X, P) = \frac{1}{\pi \hbar} \int_{-\infty}^{\infty} \Psi^{\star}(X + x) \Psi(X - x) e^{2i \frac{P.x}{\hbar}} dx.$$

For EM field, for example, we define $X = (\hat{a} e^{-i\phi} + \hat{a}^\dagger e^{i\phi})/2$ and $P = -i(\hat{a} e^{-i\phi} - \hat{a}^\dagger e^{i\phi})/2$, where ϕ is the quadrature angle. For Gaussian states where noise has a Gaussian distribution, we can write the Wigner function in terms of variance of the two quadratures that are experimentally measurable, i.e.

$$V_X = (\delta X)^2$$
$$V_P = (\delta P)^2,$$

where δ represent the standard deviation, and coherent amplitudes along X and P: δ_x and δ_p, when $\phi = 0$ [Jeong *et al.* (2007)] as

$$W(\alpha_x, \alpha_p) = \frac{2}{\pi \sqrt{V_X V_P}} e^{[-\frac{2}{V_X}(\alpha_x - \delta_x)^2 - \frac{2}{V_P}(\alpha_p - \delta_p)^2]}.$$

For a coherent state, we have $V_X = V_P = 1$ (normalized to vacuum noise) and for a pure squeezed state we have $V_X \neq V_P$ and $V_X . V_P = 1$. The following figure shows the Wigner function for a coherent state and a displaced squeezed state. The "ball-on-stick" representations are also shown in the $X - P$ planes.

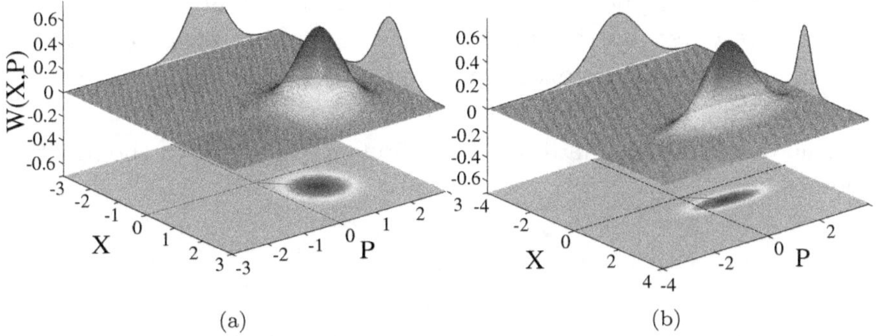

Wigner representation of (a) a coherent state (b) a squeezed state, (c) a superposition of two coherent states. The projection of the Wigner function in the x-p plane shows the size of the uncertainty in the two quadratures.

Quantum fidelity can be expressed in terms of the overlap between the ideal and measured Wigner distributions

$$\mathcal{F} = \pi \int d^2\alpha \; W_i(\alpha) W_f(\alpha).$$

Besides the Wigner representation, there are other quasi-probability functions such as the Glauber-Sudarshan P-function [Sudarshan (1963)] and Q function [Husimi (1940)].

The quantum state of a system can be accurately described if we can measure all possible outcomes. In practice, however, all measurements are incomplete. In this scenario, one can still obtain a full, but approximate, quantum description of the system using iterative methods such as the maximum likelihood estimation [Hradil (1997)]. In the maximum likelihood estimation method, we perform a set of measurements on various known quantum states (when ϕ is known) and then estimate the unknown measurements from the collected data.

Reconstruction here means the normalization of incomplete measurements that are performed on a subspace of the system. The measurements project the state to certain projectors that are mutually orthogonal. In other words, measurement projectors reproduce the identity operator $(\hat{R})$. This approach handles noisy data corresponding to realistic and incomplete measurements but with finite resolution. One can determine what are quantum states most likely represented by the measurement. This type of reconstruction can be used to estimate the quantum state of a system using, for example, homodyne measurement of quadrature values.

The detection of quadrature components reproduces the identity operator as

$$\hat{R} = \sum_i \hat{\Pi}(x_i)$$

where x_i denotes the position (or amplitude of EM field) in a particular time bin and

$$\hat{\Pi}(x_i|\phi) = \Delta x \sum_{n=0}^{\infty} \sum_{m=0}^{n} \phi_{nm}(x_i)\eta^m (1-\eta)^{n-m} \frac{n!}{m!(n-m)!} |n\rangle\langle n|,$$

$$\phi_{nm}(x|\phi) = \frac{1}{2^{(m+n)}n!m!\sqrt{\pi}} e^{-i(n-m)\phi} e^{-x^2} H_m^2(x) H_n^2(x).$$

Here η denotes the detection efficiency and H_m is the Chebyshev-Hermite polynomial of *mth* order. By analogy to the ordinary binomial distribution, we may call the random number m the number of successes in a series of n independent experiments. The reconstruction of ρ can be done using the iterative method solving the nonlinear equation for the density matrix

$$\hat{R}(\hat{\rho})\hat{\rho} = \hat{\rho}.$$

In this derivation, the condition of normalization, $Tr[\hat{\rho}] = 1$, is used such that

$$\hat{R} = \sum_i \frac{f_i}{\rho_{ii}} \hat{\Pi}(x_i|\theta)$$

$$\rho_{ii} = Tr[\hat{\rho}\hat{\Pi}(\hat{x}_i|\theta)]$$

where f_i is the probability amplitude of measuring quadrature value x_i. This method is routinely used to reconstruct the density matrix elements from the experimental quadrature data.

2.2.　Optical Resonators

Engineering resonances in different systems to increase interactions with external stimuli is ubiquitous in various fields of science and engineering. In electronics, for example, resonant circuits are used to impedance match a device to radio frequency signals for efficient communication. In particular, for quantum EM fields whose interaction with matter is typically weak and is more sensitive to losses compared to classical signals, engineering optical resonances can enhance its interaction with matter and thus reduce losses. Such efficient interaction is important in quantum systems to coherently control the quantum state of the optical fields or matter qubits.

In this section, we discuss how light can be trapped inside an optical resonator. We highlight the properties of optical resonators and their applications in enhancing light–matter interactions.

2.2.1.　*Fabry–Perot Optical Resonators*

In the simplest form, two mirrors with high reflectivity and low loss can form an optical resonator when placed along the optical propagation axis, as shown above. Optical resonators or cavities created by two reflective mirrors are called Fabry–Perot cavities. In resonant conditions, where the separation between the two mirrors is an integer number of half wavelength, light can be coupled to the resonator. If the intra-cavity medium is not a vacuum, the resonant condition becomes $L = m\lambda/2n$ where n is the refractive index and m is an integer.

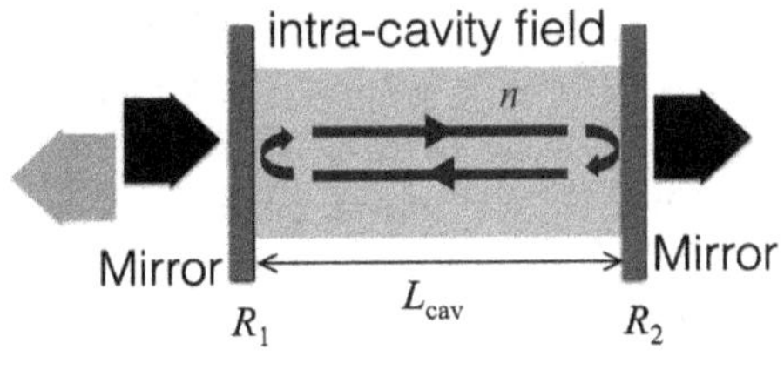

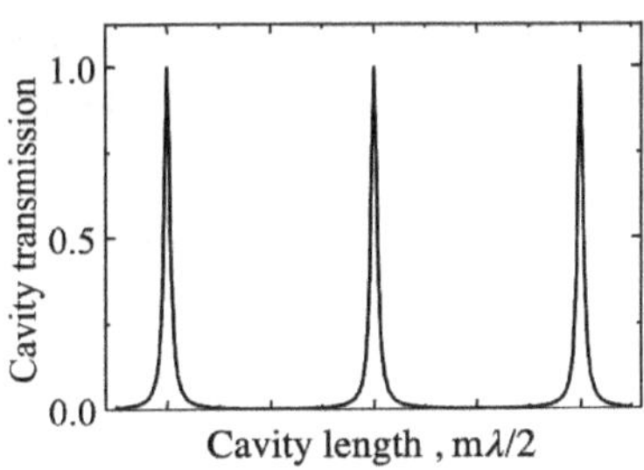

Thus the resonance condition, where maximum field is coupled to the cavity, occurs when the cavity length is equal to an integer number of half wavelengths:

$$L_{\mathrm{cav}} = m\lambda/2.$$

Once the light is coupled to the medium, it will bounce back and forth between the two mirrors, eventually leaking out of the cavity. The number of roundtrips the light takes before leaking out of the cavity depends on the mirrors' reflectivity, as shown in the equation above. To characterize the number of roundtrips and mirror reflectivity (R_1 and R_2), the quantity "finesse" ($\mathcal{F}$) is defined, which is π times the number of roundtrips.

The number of roundtrips is given in terms of **Finesse**.

$$n = \frac{\mathcal{F}}{\pi} = \frac{\sqrt{R}}{1-R}, \quad \text{for } R_1 = R_2$$

$$\text{or alternatively } \mathcal{F} = \frac{\nu_{\mathrm{FSR}}}{\nu_{\mathrm{FWHM}}}.$$

In the equation above ν_{FSR} is the frequency difference between two cavity resonances or free-spectral range, and ν_{FWHM} is the frequency bandwidth or full-width-half-maximum of the cavity resonance. Multiple resonances can be seen as the resonator length changes by $\lambda/2n_r$ from the initial resonance condition, where n_r is the intra-cavity refractive index. Equivalently, if the laser wavelength changes by $\pm\lambda/m$, additional resonances can be observed. The separation between the two resonances is referred to as the free-spectral range (FSR) measured in units of frequency or length. Also, the cavity finesse is a unitless quantity that is defined as the ratio of FSR to the full width half maximum of the cavity resonance both expressed in either frequency or length units.

The reflectivity of the mirrors R_1 and R_2 determine the efficiency of coupling light in and out of the cavity. When mirrors have equal reflectivity and losses are small, close to 100% of the light can be transmitted through the cavity at cavity resonance (i.e. critical coupling conditions). If $R_2 \gg R_1$, most of the intra-cavity light will leak through the first mirror.

2.2.2. *Examples of Optical Resonators*

Besides finesse, other resonator parameters such as quality (Q) factor, decay rate, and mode volume are also used to characterize cavity properties. The Q factor is the ratio of light energy at the resonant frequency to the dissipated energy. This quantity can be expressed as the ratio of the light frequency to the cavity linewidth in frequency units. The linewidth of the cavity resonance, when measured in units of frequency, indicates the rate of leakage of the intra-cavity light. The mode volume is defined as the volume of the intra-cavity light.

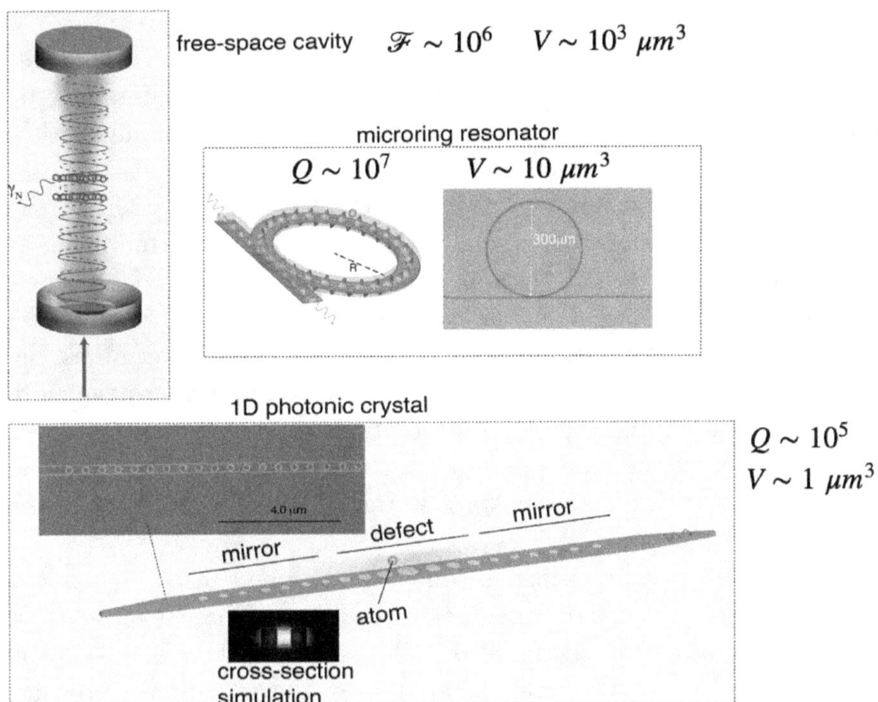

Some of these cavity parameters are defined below, including finesse:

$$\mathscr{F} = \frac{\pi \sqrt{R}}{1 - R}$$

Quality factor:

$$Q = \frac{\omega}{\Delta\omega}$$

Resonator decay rate:

$$\Delta\omega = 2\pi\left(\tau_{\mathrm{cav}}\right)^{-1} = \kappa$$

Mode volume:

$$V$$

In the case of a free-space cavity like a Fabry–Perot cavity built using two parallel curved mirrors, the intra-cavity light can be confined in the transverse direction reducing the mode volume that the EM field is confined to.

These cavity parameters are important when considering the interaction of intra-cavity light with intra-cavity matter such as atoms. Light–matter interaction is key to the generation and control of quantum optical information with applications in quantum communication. The strength of such interactions determines the efficiency and noise of the process. Optical resonators are typically used to enhance light–matter interactions. Both the Q factor or finesse as well as the mode volume affect the strength of interactions. The larger the Q factor (or smaller the volume), the stronger the interaction.

Different types of optical resonators are being used to explore different regimes of light–matter interactions. Three examples of optical resonators are shown above. The intra-cavity matter can be trapped atoms, for example, located near the maximum intensity of the intra-cavity light for maximum interaction. A micro-ring resonator, for example, guides the light around a circular path that is coupled to and from outside via a bus waveguide evanescently coupled to the mirroring. A nano-photonic cavity possesses a very small mode volume where arrays of holes in a nanobeam act like a mirror (Bragg reflector) confining light in three dimensions.

2.2.3. *Application of EM Resonators in Quantum Technology*

In practice, much stronger interaction can be achieved in microwave circuits compared to optical ones ($\lambda_{\mu w} \gg \lambda_{\text{op}}$).

In many quantum applications, the interaction of quantum light (e.g. single photons) and single atoms (or artificial atoms) is key in controlling quantum information. This is the case for both quantum computation and low-energy communication. The interaction strength between a single atom and free-space light is typically negligible. This can be understood by considering the cross-section of the light beam and the size of the atom. The minimum cross-section of a light beam is limited by the diffraction limit to about $\lambda/2.8$ which is on the order of hundreds of nanometers for visible light, while the size of the atom is on the order of a tenth of a nanometer. This suggests that a photon confined by free-space optics will most of the time pass by the atom without noticing (interacting with) the atom. Therefore, electromagnetic resonators can be implemented with high-quality factor (or finesse) and low mode volume to increase quantum interactions. For example, an optical resonator with the finesse of 3×10^6 can effectively support about 10^6 roundtrips for the photon-enhancing light–atom interaction by a factor of one million compared to propagating light without a resonator (assuming the same mode volume).

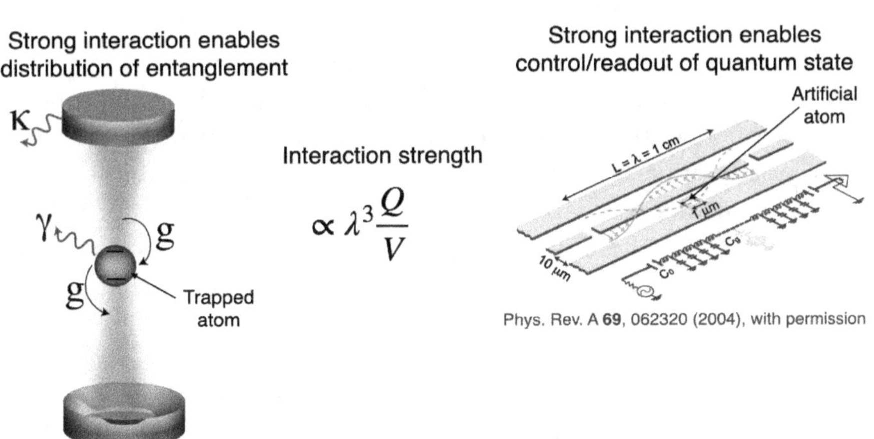

Phys. Rev. A **69**, 062320 (2004), with permission

How often a photon interacts with a single atom placed in its path is defined by coupling rate g that is inversely proportional to $\sqrt{V}$. The interaction can result in transferring a two-level atom from a low-energy state to another single state or superposition state. The atom remains in the new state for a finite time as it decays with rate γ back to the initial state. Inside a resonator, light also decays out of the resonator with the rate κ. It can be shown for a single atom to effectively interact with a single intra-cavity photon, the rate of coupling to the rate of losses needs to be close or larger than one, i.e. $g^2/\kappa\gamma > 1$. As we shall see later, this quantity is called cooperativity.

It can be shown that in certain limits, the cooperativity can purely be expressed in terms of geometrical factors, the Q factor (or finesse) and mode volume. The ratio of $\lambda^3 Q/V$ is therefore important in determining the strength of the interaction between light and atoms inside a resonator. In optical cavities, the mode volume is primarily limited by the diffraction limit, therefore it cannot become smaller. Typical cooperativities observed in optical cavity systems are on the order of 10. In contrast, in microwave cavities used for superconducting quantum circuits, a much higher ratio of $\lambda^3 Q/V$ can be obtained. Although the mode volume of microwave cavities is larger than that of most optical cavities, the wavelength of a radio-frequency EM field is much larger than the optical wavelength.

2.3. Photon Correlation

When dealing with quantum optical information, we need to perform measurements and analysis which characterize the quantum property of the light generated by a source. It is only then that we can identify whether an optical state is suitable for carrying quantum information or not. One way to characterize the quantum properties of light is to use correlation functions to quantify the degrees of correlations. The first-order correlation function or $g^{(1)}$ is used to characterize the wave nature of fields by measuring the coherence between two fields. Michelson interferometer is an example of a measurement technique used to evaluate the visibility of interference that is related to $|g^{(1)}|$, e.g. $V = 2\sqrt{I_1 I_2}/(I_1 I_2).|g^{(1)}|$, where I_1 and I_2 are intensities of EM fields (generated from a laser light split in two before the interferometer). The first-order correlation describes the frequency or phase stability of a field. For an ideal single-frequency light, we have $g^{(1)}(\tau) = e^{-i\omega\tau}$, which has a constant magnitude of 1 at all times. For a chaotic light, on the other hand, whose frequency changes over time, $g^{(1)}(t)$ has a width (decaying to zero) that signifies the timescale over which the phase of the light remains well defined.

Instead, the second-order correlation function can be used to describe the particle nature of the light field. In this section, we discuss the classical and quantum second-order correlation functions in the context of quantum optical measurements.

2.3.1. *Second-Order Correlation Function*

The second-order correlation function $g^{(2)}$ quantifies the degree of fluctuation in the intensity or photon number.

For classical fields

$$g^{(2)}(\tau) = \frac{\langle I(t)I(t+\tau)\rangle}{\langle I(t)\rangle^2}.$$

for $\tau = 0$

$$Var(I) = \langle I(t)^2\rangle - \langle I(t)\rangle^2 \geq 0$$

$$\rightarrow g^{(2)}(0) \geq 1.$$

The equation for $g^{(2)}$ above is written for the intensity of laser light as a classical measure of normalized intensity fluctuation. This can be measured experimentally using a 50/50 beam splitter and a pair of detectors, as shown in the image below.

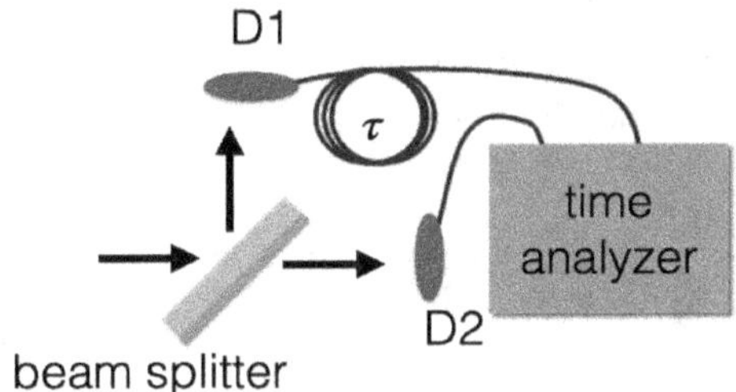

After a light beam is split in half by the beam splitter, each part is detected by a separate detector. A variable delay, τ, can be applied digitally to the detected signal. In this way, the $g^{(2)}$ will be a function of the time delay τ, which can be experimentally measured. For $\tau = 0$, the numerator in $g^{(2)}$ becomes $\langle I^2 \rangle - \langle I \rangle^2$ which is the variance of intensity fluctuation and is always positive. Therefore, we conclude that for any classical light, the $g^{(2)}(\tau = 0)$ is always greater than or equal to 1. The shape of $g^{(2)}(t)$ depends on the type of light source and its governing processes. When there is no correlation in intensity fluctuations, the $g^{(2)}$ value is equal to unity at all times. This is the case for coherent states of light (e.g. an ideal laser light) where intensity fluctuations are completely random. In other words, if we split the laser light in half, measuring the intensity of one half of the beam, it does not provide better-than-shot-noise information about the intensity of the other beam.

As an example, the intensity of a classical monochromatic light is given by $I(t) = I_0(1 + A\sin(\omega t))$. The cross-correlation function at $\tau = 0$ can be written as:

$$g^{(2)}(\tau = 0) = \frac{\langle I(t)^2 \rangle}{\langle I(t)^2 \rangle} = \frac{\langle I_0^2(1 + A\sin(\omega t))^2 \rangle}{I_0^2} = \langle (1 + A\sin(\omega t))^2 \rangle$$

using

$$\langle \sin(\omega t) \rangle = \frac{1}{T} \int_0^T \sin(\omega t) dt = 0$$

$$g^{(2)}(0) = \frac{1}{T} \int_0^T (1 + A\sin(\omega t))^2 dt = 1 + A^2/2 \geq 1.$$

An optical signal can have classical correlations. In this case, by dividing the signal in half and measuring one half, we can obtain information about the intensity of the other half. Consider an amplitude-modulated signal given by $I(t)$ above. The mean value of the intensity is I_0, but the signal has an intensity fluctuation with the peak-to-peak variation of $2A$. By splitting the signal in half and measuring both on separate detectors, we will see that the measured intensity shows in-phase oscillations around the mean intensity on the two detectors suggesting intensity correlations. We note that here we are considering a classical system where amplitude modulation A is much larger than the vacuum fluctuation and also we are ignoring other sources of noise.

Using the equation provided for $I(t)$, we can now explicitly calculate the second-order correlation function as shown above. Note that the average of the signal is given by the time integral of the signal over one period of oscillation, T, divided by T. For simplicity, we can calculate $g^{(2)}(\tau = 0)$ which will be ≥ 1.

To go from the classical description of $g^{(2)}$ to its quantum description, we need to identify the dynamical variables in the classical picture and replace them with their corresponding quantum operators. The intensity of an EM field is proportional to the photon number of the field. Thus, we can rewrite the $g^{(2)}$ equation in terms of the photon number measured at time t and $t + \tau$. The photon number has a corresponding operator in the quantum picture which can be placed into the $g^{(2)}$ equation to derive the quantum equation for the correlation function.

We can rewrite the correlation function for a classical field, replacing intensity with field squared and then write field amplitudes in terms of ladder and number operators:

$$g^{(2)}(\tau) = \frac{\langle I(t)I(t + \tau)\rangle}{\langle I(t)\rangle^2}$$

by replacing $I(t) \propto n(t) \propto E(t)^2 n \to \hat{n} = \hat{a}^\dagger \hat{a}$, we arrive at

$$g^{(2)}(\tau) = \frac{\langle n(t)n(t + \tau)\rangle}{\langle n(t)\rangle^2}.$$

In the quantum picture, and using Wick's theorem, we can write the quantum form of the 2nd order correlatipon function as

$$g^{(2)}(\tau) = \frac{\langle \hat{a}^\dagger(t)\hat{a}^\dagger(t + \tau)\hat{a}(t + \tau)\hat{a}(t)\rangle}{\langle \hat{a}^\dagger(t)\hat{a}(t)\rangle^2}.$$

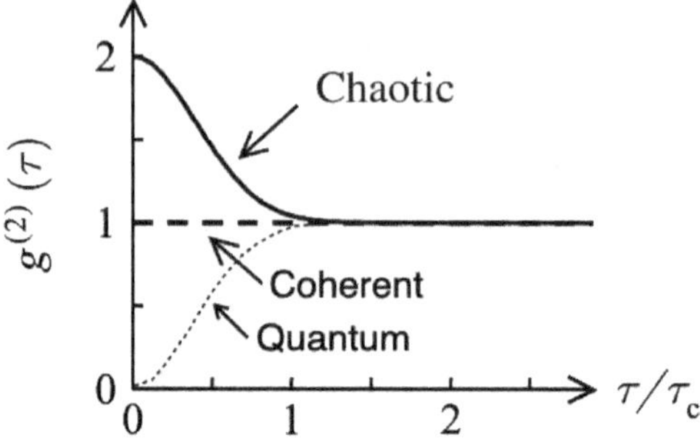

Note that when you derive the quantum equation from its classical counterpart, there is some freedom in the operator order. We can have all creation operators to the left (normal order) or the right (anti-normal order). The choice of operator order yields different ground state energies. In the above equation, when replacing $\hat{n}$ with $\hat{a}^\dagger \hat{a}$, the denominator becomes $\langle \hat{a}^\dagger(t)\hat{a}(t)\hat{a}^\dagger(t+\tau)\hat{a}(t+\tau)\rangle$. Using normal ordering of the operators or Wick ordering (i.e. all creation operators are to the left of all annihilation operators in the product), we can rewrite the numerator as $\langle \hat{a}^\dagger(t)\hat{a}^\dagger(t+\tau)\hat{a}(t+\tau)\hat{a}(t)\rangle$. Also, note that $\langle \cdots \rangle$ indicates the expectation value of the operator enclosed by the bra–ket. If the form of the quantum state in the Fock basis (eigenstates of the number or energy operator) is known, the second-order correlation can be calculated. For example, a single-photon state in the Fock basis can be expressed as $|1\rangle$. So, the numerator of the $g^{(2)}(\tau = 0)$ becomes $\langle 1|\hat{a}^\dagger(t)\hat{a}^\dagger(t)\hat{a}(t)\hat{a}(t)|1\rangle$.

Using the eigenvalue equation for the annihilation operator, $\hat{a}|n\rangle = \sqrt{n}|n-1\rangle$, we find that $\hat{a}(t)\hat{a}(t)|1\rangle = 0$, resulting in a zero-valued correlation function. We see that for such a simple quantum state, i.e. a single-photon state, the correlation function takes a value not allowed by classical statistics. In general, the correlation function can take a range of values shown in the graph above where time τ is normalized to τ_c, the effective coherence time beyond which quantum correlations can not be observed.

For light with classical correlation (chaotic or thermal light) $g^{(2)}(\tau) \geq 1$ for non-correlated light (coherent states) $g^{(2)}(\tau) = 1$ and quantum correlated light $g^{(2)}(\tau) \leq 1$. In both classical and quantum correlations, measuring half of a beam provides information about the other half. The difference is in the level of uncertainty in predicting the value of the unmeasured signal. For classical light, the best accuracy will be worse than shot noise. For quantum correlation, our prediction can be made with sub-shot noise precision.

Another way to interpret the meaning of the value of the cross-correlation function is in terms of bunching and anti-bunching of photons. For a classical light with some intensity modulation, photons arrive in bunches. Detection of a photon in a portion of the beam signals the presence of more photons in other portions or segments, and vice versa. In the case of a single-photon quantum state, however, detecting a single photon in a portion of the beam suggests a close-to-zero possibility of detecting a second photon in other segments during the same time window, i.e. anti-bunching. For a coherent state where the photon fluctuations around the mean value are governed by a random process, photons neither bunch nor anti-bunch. The photon number, in this case, varies randomly at every shot.

2.3.2. *Hong–Ou–Mandel Interference*

We can write the quantum state of the two photons after the beam splitter using the creation operators applied to a vacuum state. Before photons arrive at the beam splitter, the state after the beam splitter is a vacuum state. When photon 1 is traveling along the horizontal direction (mode a) and photon 2 is traveling along the vertical direction (mode b) and pass through the beam splitter, multiple scenarios can occur. Firstly, photon 1 can be transmitted while photon 2 is reflected. This possibility is described by the term $a_1^\dagger a_2^\dagger |0\rangle$ where both photons appear in mode a. When both photons are reflected, photon 1 appears in mode b and photon 2 appears in mode a. This possibility is represented by the term $a_2^\dagger b_1^\dagger |0\rangle$. Similarly, we can write the two other possibilities; however, as can be seen, the other probabilities have a negative sign. The sign is related to the relative phase of the two photons when they are reflected or transmitted and can be classically explained by the refractive index contrast in the beam splitter. The relative phase of the fields changes by 180 degrees when reflected from a surface with a higher refractive index. The coating in the glass slide that gives a beam splitter its functionality has a refractive index lower than the glass but higher than air. The asymmetry in the index difference for the two photons gives rise to the sign change in the equation above. As a result, two of the possibilities (probability amplitudes), i.e. one photon in each path, destructively interfere and are canceled out. For a more formal treatment of a quantum beam splitter and derivation of the above output state, we refer the reader to book *Introductory Quantum Optics* by C.C. Gerry and P.L. Knight (Chapter 6).

Therefore, when two indistinguishable photons interfere on a beam splitter, the probability amplitudes for "both transmitted" and "both reflected" events are perfectly canceled out:

$$|\psi_{12}\rangle = C(a_1^\dagger a_2^\dagger - a_2^\dagger b_1^\dagger + a_1^\dagger b_2^\dagger - b_1^\dagger b_2^\dagger)|00\rangle$$

The result of this interference is the bunching of the two photons in mode a or b. The negative sign is also a mathematical requirement to ensure the reversibility of the beam splitter operation. The indistinguishability of these photons suggests that we cannot separate the two possibilities (one photon in each path) and they are canceled out. The coefficient C is the normalization coefficient that is equal to $1/2$. For the interference to occur, the two photons need to have the same frequency and polarization. They also need to temporally and spatially overlap on the beam splitter, hence the word indistinguishable.

An experiment devised by Hong, Ou, and Mandel in Rochester (1987) demonstrated that when two identical photons interfere in a beam splitter, the photons tend to bunch after leaving the beam splitter.

2.4. Atoms

Most quantum technologies rely on atoms (or artificial atoms) and photons as their building blocks to generate, store and process quantum information. Photons of different wavelengths can be treated as flying qubits carrying quantum information. Atoms on the other hand are stationary qubits that also hold quantum information. The interaction between atoms and photons is used to control and manipulate the state of the qubits. In many cases, the interaction between two stationary qubits can be mediated by a flying qubit and vice versa. Atomic-based qubits are used to implement quantum processors and memories while relying on interaction with optical photons for control and processing. On the other hand, superconducting circuits, used as a platform for quantum computing, rely on engineering techniques to build artificial atoms using electronic circuits interacting with photons at radio frequency. Therefore, understanding atomic properties and light–atom interaction in the quantum regime is the key to the implementation of many quantum applications.

In this section, we provide a basic description of atoms in the quantum regime and discuss different ways atoms can interact with photons.

2.4.1. *Quantization of Energy in Atoms*

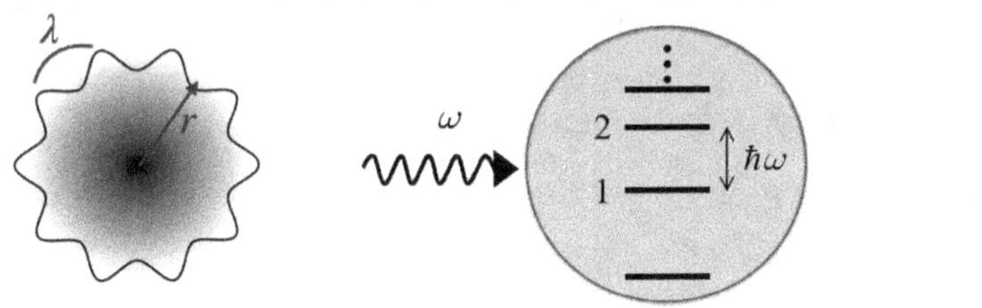

Allowed orbit   Hydrogen-atom energies   Isolated two energy levels

$$2\pi r = (n+1)\lambda \qquad E_n \propto \frac{1}{n^2} \qquad |\psi\rangle = c_1|1\rangle + c_2|2\rangle$$

Louis de Broglie in 1924 postulated that if waves can have particle-like properties, e.g. quantization of EM field energy, particles should also have wave-like properties. He associated a wavelength, λ to a particle with momentum p given by $\lambda = h/p$, where h is the Planck constant. Bohr in 1913 suggested that electrons oscillating around atoms are separated from the nucleus by a distance given by the probable radius (Bohr radius, r). If now the electrons are treated as waves, for electrons to have an allowed circular path around the atoms, the circumference of the path should be an integer multiple of the Bohr wavelength. From this picture, one can calculate the energy of the electron using the Coulomb potential. For a simple atom like Hydrogen, it can be shown that the electron's energies are quantized and inversely proportional to an integer n. However, unlike the harmonic oscillator, the energy spacing between atomic energy levels is not constant and varies with n. This makes an atom a "nonlinear oscillator" that enables it to interact with EM fields. For example, if the energy spacing between two arbitrary energy states of an atom, $E_2 - E_1$, is equal to the energy of an incident photon, $\hbar\omega$, the photon can uniquely address $E_1 \to E_2$ transition of the atom. If we restrict the system to this energy range, the atom can be simplified such that only two energy states of the atom need to be considered. This creates a stationary qubit (a two-level quantum system) that interacts with the flying qubit. As long as an atom has favorable coherence properties and can efficiently interact with photons, the atom–photon system can be exploited for various quantum applications. By controlling the interaction strength, light intensity, and duration, an atom initially prepared in state $|1\rangle$ can be transferred to any arbitrary superposition of two states $|1\rangle$ and $|2\rangle$.

2.4.2. *Two-Level Atom and Bloch Sphere*

Considering only the two energy levels of an isolated atom, we can write the wave function of the atom on an energy basis using the probability amplitudes c_1 and c_2. Although energies are discrete, the probability amplitudes are complex numbers that belong to a continuum. A two-level qubit, therefore, can be found in an infinite number of possible states. To show the continuity in the wave function, we can represent the coefficients, c_1 and c_2, using generic angles θ and ϕ. These angles construct a three-dimensional spherical coordinate where a vector can be drawn from the origin to a point in this space to represent a wave vector or a wave function. As the length of the vector (along x, y, z dimensions) has to be equal to one, all possible points form a sphere, which we refer to as the Bloch sphere, and each wave vector is the Bloch vector in this sphere. We can write

$$|\psi\rangle = \cos(\theta/2)|2\rangle + \sin(\theta/2)e^{i\phi}|1\rangle$$

$$|\psi\rangle = c_1|1\rangle + c_2|2\rangle$$

$$|c_1|^2 + |c_2|^2 = 1$$

This geometrical representation of the wave vector of a qubit enables the visualization of the wave function's evolution after various operations, where

$$x = \sin\theta\cos\phi$$

$$y = \sin\theta\sin\phi$$

$$z = \cos\theta$$

Two-level atom

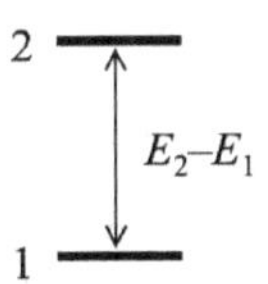

Bloch Sphere

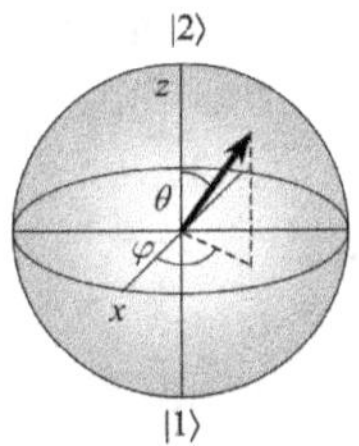

Once an atom is prepared in a specific combination of two energy levels, its state can undergo decay and random fluctuations due to interactions with the environment. Damping or decay processes are typically characterized by two-time constants T_1 and T_2. The T_1 time refers to the decay time of the population. The decay of the population can occur through spontaneous radiative decay or non-radiative processes. The T_2 time is a measure of dephasing or decoherence. Fluctuation of energy levels through collision or field fluctuations may cause the phase of the wave function to vary causing coherence (or the "waveness of the particle") to vanish even without population decay. Note that states $|1\rangle$ and $|2\rangle$ represent eigenstates of number or energy (Hamiltonian) operator that we also refer to as Fock states that are number states with a well-defined number of quanta.

2.4.3. *Atomic Operators*

Instead of states 1 and 2, we can use the labels g and e for the ground and excited states of energy.

$$|\psi\rangle = c_1|1\rangle + c_2|2\rangle$$

$$|\psi\rangle = c_{\mathrm{g}}|g\rangle + c_{\mathrm{e}}|e\rangle$$

Similar to EM fields, we replace the dynamical variables with operators, e.g. Pauli spin operators.

An operation on an atomic qubit maps the initial two-dimensional wave function to another two-dimensional wave function representing the qubit. This is equivalent to the rotation of the Bloch vector. The commonly used atomic operators are Pauli spin operators that can rotate the Bloch vector around the three axes of the Bloch sphere. The operators are 2×2 matrices given by

$$\hat{\sigma}_x = \begin{pmatrix} 0 & 1 \\ 1 & 0 \end{pmatrix}, \quad \hat{\sigma}_y = \begin{pmatrix} 0 & -i \\ i & 0 \end{pmatrix}, \quad \text{and} \quad \hat{\sigma}_z = \begin{pmatrix} 1 & 0 \\ 0 & -1 \end{pmatrix}$$

$$\hat{\sigma}_{xyz} = \hat{\sigma}_x \bar{i} + \hat{\sigma}_y \bar{j} + \hat{\sigma}_z \bar{k}.$$

Using these operators, we can define the rotation operators for a Bloch vector along the x, y, or z axis as

$$R_x(\theta) = e^{-i\theta\sigma_x/2}, \quad R_y(\theta) = e^{-i\theta\sigma_y/2}, \quad R_z(\theta) = e^{-i\theta\sigma_z/2}.$$

We can also define the ladder operators for atomic states, which are analogous to ladder operators of EM fields.

$$\hat{\sigma}^{\dagger} = \hat{\sigma}_x + i\hat{\sigma}_y = |e\rangle\langle g| \quad (\text{similar to } a^{\dagger})$$

$$\hat{\sigma} = \hat{\sigma}_x - i\hat{\sigma}_y = |g\rangle\langle e| \quad (\text{similar to } a)$$

$$\hat{\sigma}_z = |e\rangle\langle e| - |g\rangle\langle g| \quad (\text{similar to } a^{\dagger}a)$$

For example, the operator $\sigma^{\dagger}$ when applied to an atom initially in the g state transfers the atom to the excited state, e. Also, the expectation value of the σ_z operator gives the probability of finding an atom in the excited state, in other words, the population of the atom. We can use the matrix form or the ket–bra form of the operators to find the resulting state after applying an operation. This depends on whether the state is written in vector or bra–ket form. For example, when initially $c_g = 0$,

$$\hat{\sigma}|e\rangle = \begin{pmatrix} 0 & 2 \\ 0 & 0 \end{pmatrix}\begin{pmatrix} 0 \\ 1 \end{pmatrix} = \begin{pmatrix} 2 \\ 0 \end{pmatrix}, \quad \text{or} \quad \hat{\sigma}|e\rangle = |g\rangle\langle e|e\rangle = |g\rangle.$$

Note that state $\begin{pmatrix} 2 \\ 0 \end{pmatrix}$ after normalization becomes $\begin{pmatrix} 1 \\ 0 \end{pmatrix} = |g\rangle$.

2.5. Light–Atom Interactions

Atoms as stationary qubits play an important role in quantum technology as they can store and process quantum information. Electromagnetic fields in the optical and microwave wavelengths can be used to mediate interactions between two or more atoms. In quantum communication, atoms are often used for the generation and storage of quantum optical states carrying quantum information. It is therefore important to review principles governing light–atom interactions before we discuss the application of such interactions in quantum communication.

In this section, we introduce the basic principles of light–atom interaction using a simplified model of atoms where only two to three energy levels of atoms are considered.

2.5.1. *Absorption and Emission*

An electromagnetic field can interact with atoms in different ways. An incoming photon can be absorbed by the atom if the photon energy matches the energy spacing between any two energy levels of the atom. Consider a two-level atom initially in the ground energy state $|g\rangle$. Upon absorption, the atomic state changes from $|g\rangle$ to $|e\rangle$. $|e\rangle$ is the atom's excited state.

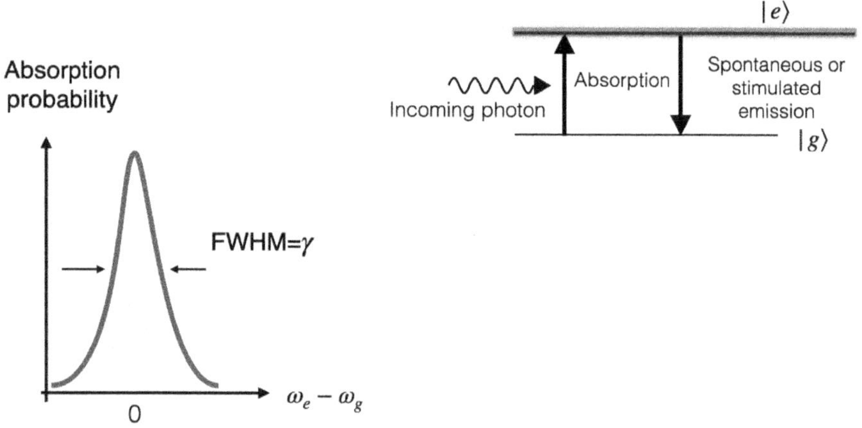

If the atom is initially in the excited state, it can decay to the lower energy state via a spontaneous or stimulated process. Spontaneous emission is governed by the interaction of the atom with the vacuum fluctuation that gives rise to the exponential decay of the atom to the lower state with a time constant equal to the inverse of the bandwidth of the emitted light. Using the Heisenberg principle, the relation between the time and energy (or frequency) uncertainty of the photon can be determined. The uncertainty in photon energy can be translated to uncertainty in atomic energy, suggesting that there is an uncertainty associated with determining the energy spacing of any two energy levels of atoms (γ). We refer to this as natural linewidth or lifetime broadening of the atoms described by the following uncertainity relation.

$$\Delta E \Delta t \geq \hbar/2$$

The atoms' energy levels can be influenced by other broadening mechanisms, such as Doppler broadening and collisional broadening. If all the atoms in the system are affected by the broadening mechanisms, in the same way, the broadening is said to be homogenous; otherwise, the atoms' spectrum is said to be inhomogeneously broadened. Lifetime and Doppler broadening are examples of homogeneous and inhomogeneous broadening mechanisms, respectively. The absorption or emission lineshape for a homogeneously (inhomogeneously) broadened system has a Lorentzian (Gaussian) profile.

In the stimulated emission process, the excited atom interacts with external resonant stimuli (incident EM field). The external field forces the atom from $|e\rangle$ to $|g\rangle$ by causing the atom to emit a photon with the same phase as the incident field (coherent emission). Lasers rely on the stimulated emission to create a strong and coherent EM field by placing an ensemble of atoms inside an optical cavity.

2.5.2. *Atomic Dipole*

Now, let us discuss the physics that governs light–atom interactions. The classical electromagnetic theory describes EM radiation and absorption by an oscillating dipole. When electrons are unevenly distributed around the nucleus, an electric dipole can be created where two clouds of charged particles (electrons and protons) are separated by a distance, r. For a single electron, the dipole is given by $d = q.r = -er$. The classical Hamiltonian or interaction energy of the dipole is then calculated by calculating the electric potential energy as given above.

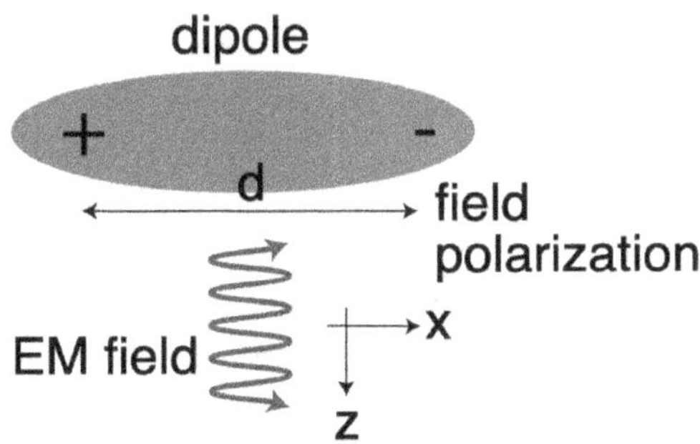

The interaction energy between electron and E-field:

$$H = F \cdot r = d \cdot E$$

where $\mathbf{r}$ is the electron distance from the nucleus (electron orbit).

$$d = -e\mathbf{r}.$$

In this case, the maximum absorption or emission occurs when EM field polarization is aligned with the dipole direction and EM frequency matches the frequency of the oscillating dipole.

The electric field with polarization along the dipole axis propagating along the z direction, as shown in the diagram above, can be written as $E = E_0\hat{x}\cos(kz - \omega t)$. Frequency ω is equal to the frequency difference between energy levels of the atom. The radiation or absorption from the two energy levels of an atom can occur only if the dipole transition is allowed. If the electron's orbits (shape of the electron path or trajectory) corresponding to two energy levels are the same, the relative electron distribution does not change as a result of that transition, therefore no dipole is created and no radiation is emitted. The distance r is the change in the electron orbital distance when going between two energy levels and can be expressed in terms of momentum as $r = p_x/m\omega$, where p_x is momentum and m is the effective electron mass. Selection rules (set of restrictions associated with the transition of an atoms, e.g. transition is allowed if it leads to the orbital quantum number of an electron changing by one and its magnetic quantum number remains the same or changes by one) such as momentum selection rules determine whether an electric–dipole transition is allowed or forbidden.

2.5.3. *Light–Atom Interaction*

To write the quantum Hamiltonian corresponding to the interaction energy of the system, we replace the dipole d with its equivalent operator form in quantum mechanics.

Starting from the classical interaction Hamiltonian or potential,

$$H = F \cdot r = d \cdot E$$

and replacing dynamical variables with operators, we arrive at the quantum Hamiltonian describing light–atom interactions.

$$d \to \hat{d} = \mu(\hat{\sigma} + \hat{\sigma}^\dagger)$$

where we define

$$\mu = \langle g|er|e \rangle \equiv d \quad \text{dipole moment}$$

$$\Omega = \frac{|\mu \cdot E|}{\hbar} \qquad \text{Rabi frequency}$$

The expectation value of d is shown by μ. The atomic operators σ and $\sigma^\dagger$ are defined in terms of Pauli matrices and describe electron transition between two quantized energy levels of the atoms. Above, we also define a quantity known as the Rabi frequency that represents the magnitude of the light–atom interaction.

The dipole moment μ depends on the particular transition in an atom as well as the polarization of the EM field. For Rubidium atoms, for example, the effective dipole moment is on the order of 10^{-29} C.m multiplied by a factor taking into account the momentum dependency. This factor is known as the Clebsch–Gordan coefficient and accounts for the momentum or polarization of the EM field as well as the momentum and orbit of the electron in the corresponding energy levels. The Clebsch–Gordan coefficients range from 0 to 1.

2.5.4. *Light–Atom Hamiltonian*

To write the total Hamiltonian for a single atom interacting with a single photon, we need to add the energy of the photon and the atom to the total interaction energy. The Hamiltonian of an EM field can be written in terms of ladder operators, as shown before. The energy of the atom can be written in terms of σ_z which is equivalent to the number operator for the atomic field.

The total Hamiltonian of the system consists of three parts, atomic, optical and light–atom interaction Hamiltonians:

$$\hat{H}_L = \hbar\omega_0 \hat{a}^\dagger \hat{a} \quad \text{(ignoring constant 1/2)}$$

$$\hat{H}_A = \hbar\omega_{eg}\hat{\sigma}^\dagger\hat{\sigma} = \hbar\omega_{eg}(|e\rangle\langle e|)$$

$$\hat{H}_{LA} \cong \hbar g(\hat{\sigma}\hat{a}^\dagger + \hat{\sigma}^\dagger\hat{a})$$

where light–atom interaction strength is given by

$$g = \mu\sqrt{\frac{\omega_{eg}}{2\epsilon_0\hbar V}} = \Omega \quad \text{Rabi frequency} \quad \Omega = \frac{|\mu \cdot E|}{\hbar}$$

To write the interaction Hamiltonian, we need to multiply the operator forms of d and E as

$$\hat{H}_{LA} = \hat{d}.\hat{E} = \hbar g(\hat{\sigma}\hat{a}^\dagger + \hat{\sigma}^\dagger\hat{a} + \hat{\sigma}^\dagger\hat{a}^\dagger + \hat{\sigma}\hat{a})$$

where g is the Rabi frequency for a single photon, often called the light–atom coupling rate.

The last two terms in the Hamiltonian correspond to the creation of the photon when an atom is excited or the annihilation of a photon when an atom is de-excited. Both of these processes can be ignored as these processes are not likely to occur when we have only one atom and one photon. Therefore, the interaction Hamiltonian can be simplified by keeping the $\hat{\sigma}\hat{a}^\dagger$ and $\hat{\sigma}^\dagger\hat{a}$ terms.

The total Hamiltonian is then given by $H = \hat{H}_L + \hat{H}_A + \hat{H}_{LA}$, which is known as the Jaynes–Cummings (JC) Hamiltonian. It describes the interaction between a two-level atom and a quantized mode of a monochromatic optical field (e.g. a single photon defined in a single spatial mode like a cavity mode).

2.5.5. *Light–Atom Dynamics*

Using the JC Hamiltonian and time-dependent Schrödinger equation, we can now describe the basic dynamics of the light–atom interactions.

Considering a single atom initially in the ground state interacting with a single photon, the possible quantum state of atom–photon can be expressed as

$$|\psi\rangle = c_{g,1}|g\rangle|1\rangle + c_{e,0}|e\rangle|0\rangle.$$

Using the Schrödinger equation, we can now describe the evolution of a single photon–atom system:

$$\frac{\partial}{\partial t}|\psi\rangle = -\frac{i}{\hbar}\hat{H}|\psi\rangle$$

The generic pure quantum state of the light–atom system is given by $|\psi\rangle$ above, where $|g\rangle$ and $|e\rangle$ represent the ground and excited energy states of an atom and $|1\rangle$ and $|0\rangle$ represent a quantized field in the Fock basis having 0 or 1 photons. If an atom is found in $|g\rangle$ state, the photon is not absorbed; if the photon is absorbed, the atom will then be excited to $|e\rangle$. The coefficients c_{g1} and c_{e0} are the probability amplitudes of finding the system in either configuration.

We can now write the Schrödinger equation to find how these probability amplitudes change with time and how they can be controlled by changing the physical quantities. Considering the pure state defined above we can write the corresponding time-dependent Schrodinger equation to obtain equations of motions describing the dynamics of the combined light-atom state

$$|\Psi\rangle = c_{g1}|g\rangle|1\rangle + c_{e0}|e\rangle|0\rangle$$

$$\frac{\partial}{\partial t}|\Psi\rangle = -\frac{i}{\hbar}\hat{H}|\Psi\rangle.$$

On the left-hand side of the Schrödinger equation above, we take the partial time derivative of the wave function to find the dynamics of the probability amplitudes, e.g. $\partial c_{g1}/\partial t = \dot{c}_{g1}$. On the right-hand side, we need to apply the Hamiltonian operator to the wave function. The result of these operations is the following rate equation.

$$\dot{c}_{g,1}|g,1\rangle + \dot{c}_{e,0}|e,0\rangle = -i\Omega c_{g,1}|e,0\rangle - i\Omega c_{e,0}|g,1\rangle$$

$$-i\omega_{eg}c_{e,0}|e,0\rangle + 0$$

$$0 + i\omega_0 c_{g,1}|g,1\rangle.$$

To see how we can arrive at the terms on the right-hand side, we need to evaluate the operations term by term. For example, for the term corresponding to the light–atom interaction, we have

$$-\frac{i}{\hbar}\hat{H}_{LA}|\psi\rangle = -ig(\hat{\sigma}\hat{a}^\dagger + \hat{\sigma}^\dagger\hat{a})(c_{g1}|g\rangle|1\rangle + c_{e0}|e\rangle|0\rangle)$$

which consists of four terms, two of them are zero because more than one excitation is not allowed in the system (only one photon and one two-level atom) and the other two are found using

$$\hat{\sigma}\hat{a}^{\dagger}|e\rangle|0\rangle = \hat{\sigma}|e\rangle|1\rangle = |g\rangle\langle e|e\rangle|1\rangle = |g\rangle|1\rangle$$

$$\hat{\sigma}^{\dagger}\hat{a}|g\rangle|1\rangle = \hat{\sigma}^{\dagger}|g\rangle|0\rangle = |e\rangle\langle g|g\rangle|0\rangle = |e\rangle|0\rangle$$

where we first apply the field ladder operators and then the atomic operators and we use normalization conditions, $\langle e|e\rangle = 1$ and $\langle g|g\rangle = 1$. Similarly, we can find the effect of the light and atomic Hamiltonian on the wave function.

Note that $\hat{a}^{\dagger}|1\rangle = \hat{a}|0\rangle = \hat{\sigma}^{\dagger}|e\rangle = \hat{\sigma}|g\rangle = 0$ considering the simplified system above. In the case where more than two energy levels of an atom are involved and more photons and atoms are considered, additional terms need to be also included in the equations of motion.

2.5.6. *Rabi Oscillation*

We can now arrive at the two equations of motion for the probability amplitudes by considering the coefficients of $|g\rangle|1\rangle$ and $|e\rangle|0\rangle$ on the two sides of the Schrödinger equation separately. To simplify the equations, we define the detuning term, Δ, as the frequency difference between the optical and atomic frequencies.

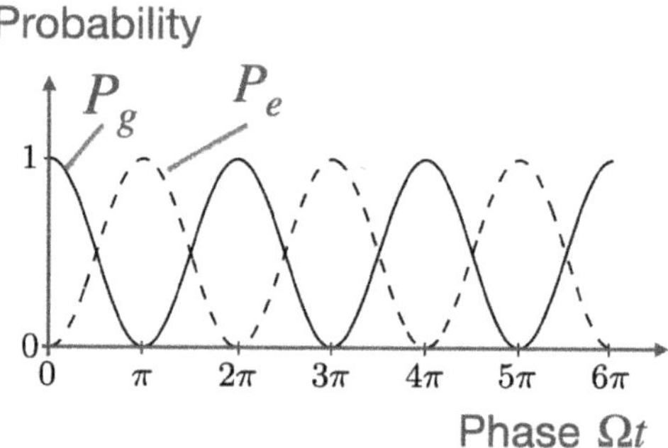

The equation of motion considering only the atomic part of the wave function is given by

$$\dot{c}_{e,0} = -i\Omega c_{g,1}$$

$$\dot{c}_{g,1} = -i\Delta c_{g,1} - i\Omega c_{e,0}$$

where detuning:

$$\Delta = \omega_{eg} - \omega_0$$

which has the general solution for $\Delta = 0$ (considering only the atomic part):

$$c_g(t) = A\sin(\Omega t) + B\cos(\Omega t)$$
$$c_e(t) = C\sin(\Omega t) + D\cos(\Omega t) \qquad \longrightarrow \qquad P_e = |c_e|^2 \qquad P_g = |c_g|^2$$

The sinusoidal solutions to these differential equations are shown above. The coefficients A–D are found using the initial conditions and continuity condition of the wave function. Assuming that at $t = 0$ the atom is in the ground state, the probabilities are plotted as a function of Ωt.

Such oscillations are called Rabi oscillations, and they indicate a coherent oscillation of the population between the two atomic energy levels when interacting with a continuous EM field. The oscillation is considered a characteristic quantum signature in isolated two-level quantum systems.

By changing the duration or Rabi frequency, the probability to find the atom in the excited state can be determined. When the EM field is pulsed with a certain duration such that $\Omega t = \pi$ or $\pi/2$, it is called π or $\pi/2$ pulses, respectively. A π pulse inverts the population and a $\pi/2$ creates a superposition state from an atom initially in $|e\rangle$ or $|g\rangle$.

After applying a π pulse and monitoring the probability of finding the particle in either state after different delays, one can determine the population decay time (T_1).

To perform measurements of dephasing or the T_2 time, we refer to a method proposed by Ramsey (Physics Nobel, 1989):

(1) First a $\pi/2$ pulse is applied to an atom, which is initially in the ground state. This flips the Bloch vector into the x–y plane.

(2) If there is no detuning (frequency shift or frequency noise) or dephasing, then a second $\pi/2$ pulse after some time t will flip the Bloch vector exactly to the excited state, which can then be detected by observing the empty $|g\rangle$ state (e.g. through lack of absorption).

(3) If, however, there is some detuning or phase fluctuation, then the Bloch vector rotates in the x–y plane by an angle δt. A second $\pi/2$ pulse would then not tilt the Bloch vector exactly to the excited state.

By measuring the fraction of the atomic population in one state for varying delays between the two $\pi/2$ pulses, one can determine the T_2 time.

2.6. Strong Light-Atom Interactions

The strength of interactions between photons and atoms is important for efficient and coherent control of quantum optical information. In introducing the Heisenberg uncertainty principle, we saw that the spatial footprint of an EM field in free space is given by the diffraction limit. The spatial footprint of photons is typically much larger than the size of an atom. This suggests that when the two are interfaced in free space, most of the time the photon will pass by the atom without interacting with it. In reverse, when an excited atom emits a photon, the photon can be collected with limited efficiency as most times the emitted photon's mode does not match the detection mode. This basic description indicates that both absorption of a photon by an atom and the collection of an emitted photon from an atom in free space occurs probabilistically. When a photon carries quantum information, this probabilistic interaction results in a loss of information in the form of, for example, a low probability of the creation of entanglement between photons or atoms.

Interaction can be enhanced by optical resonators to overcome the probabilistic limit in interactions. In this section, we discuss the strong interaction between a photon and an atom mediated by optical cavities. Such interaction is key to deterministic entanglement creation between atomic and photonic qubits, as well as superconducting qubits, and are used in various quantum technologies.

We note that simply placing an atom inside an optical cavity does not make its interaction with a photon less probabilistic. Certain interaction parameters need to be engineered to reach the strong coupling regime. We refer to this interaction regime as high cooperativity or strong coupling regime, where an atom or a photon state can interfere with itself. Consider that a photon emitted from the atom is not lost to the environment and can bounce back and interact with the atom again. If losses are small and interactions can happen faster than the decay rate of the atom, the EM field of the photon can interfere with the atom and suppress or enhance the emission. The interaction is therefore nonlinear at the single particle level. This strong nonlinearity has applications in deterministic entanglement creation.

2.6.1. *Light–Atom Interaction in Free Space*

The scattering power (or emission probability) by an atom into a particular mode or direction, P_M, relative to the scattering power in all directions (4π solid angle) is called "cooperativity". P_{in} is the incident power. The resonant free-space cooperativity, η_{fs}, can be defined in terms of geometrical factors: the wave number $k = 2\pi/\lambda$ and mode waist, w, or mode area A.

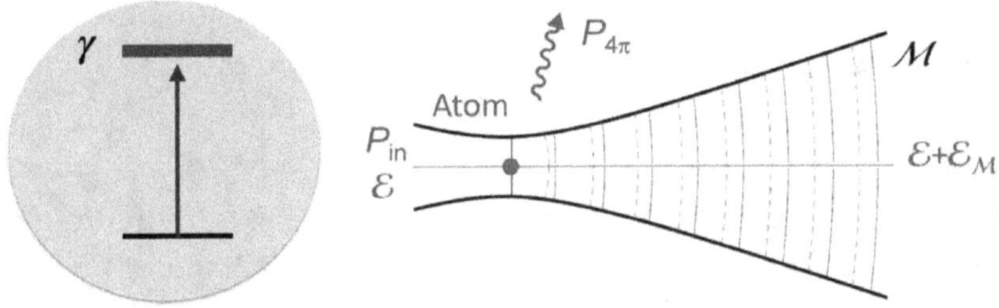

The scattered light can interfere with the incident field modifying scattering characterized by cooperativity as below:

$$\frac{P_M}{P_{4\pi}} = \eta_{\text{fs}}$$

$$\frac{P_{4\pi}}{P_{in}} \simeq 2\eta_{\text{fs}} \xrightarrow{\text{Cooperativity}} \eta_{\text{fs}} = \frac{6}{k^2 w^2}$$

where

$$k = 2\pi/\lambda \ \ \& \ \ A = \pi w^2/2.$$

Away from the atomic resonance, the free-space cooperativity is modified by a factor written in terms of the excited state atomic linewidth, γ, and light–atom detuning Δ.

$$\eta_{\text{fs}} \to \eta_{\text{fs}} \left(\frac{\gamma^2}{\gamma^2 + 4\Delta^2} \right)$$

The detailed derivation of cooperativity is provided in Tanji-Suzuki *et al.* (2011). As the wavelength of light is bounded by the atomic transition in free space, the cooperativity can only be increased by focusing the light more at the location of the atom. The assumption used in the derivation of the cooperativity relationship above is that the coherence time of the atom and photon should be longer than the interaction time given by the photon wave packet duration.

Considering the diffraction limit, $w \gtrsim \lambda/2.8$, the free-space cooperativity can, in principle, reach values on the order of unity. A cooperativity on the order of one and higher corresponds to the "strong coupling regime".

2.6.2. *Cavity Interactions*

Inside an optical resonator, the interaction can increase by a factor proportional to the cavity finesse, $\mathcal{F}$, compared to free space. The enhancement factor is determined by the number of intra-cavity roundtrips a photon takes before it leaves the cavity.

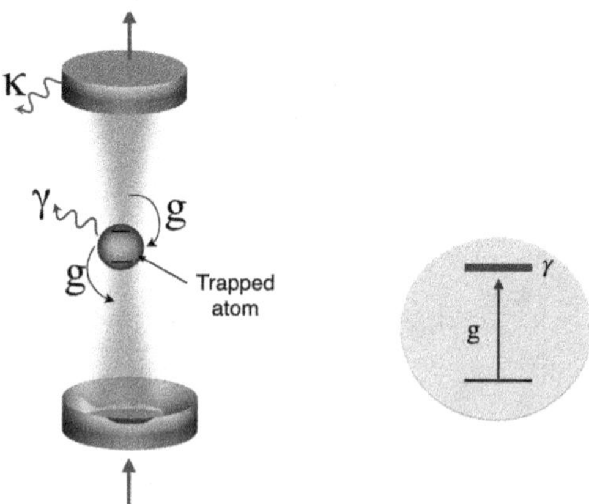

A single atom can modify light transmission. The size of such modification depends on the strength of interaction, characterized by cooperativity:

$$\eta = \mathcal{N}\eta_{\text{fs}} = \frac{24\mathscr{F}/\pi}{k^2 w^2}$$

or equivalently

$$\eta = \frac{4g^2}{\kappa\gamma}$$

where:

$$\mathcal{N} = \frac{4\mathscr{F}}{\pi} = \frac{4\sqrt{R}}{1 - R}, \quad \text{for } R_1 = R_2.$$

See Tanji-Suzuki *et al.* (2011).

Equivalently, the cooperativity is defined as the ratio of the light–atom interaction rate to the loss rates by atom, γ, and cavity, κ. The cavity finesse is given by mirror reflectivities. Cavity finesse as high as 100,000 or more is achievable with bulk mirrors. To reduce the mode volume or cross-section, w, mirrors with a large radius of curvatures are used. Alternatively, micro- or nano-photonic cavities can be fabricated to reduce the mode volume or cross-section. We note that the cooperativity above is the maximum possible cooperativity and the actual value can only be less than the maximum value due to, for example, non-ideal atom–field overlap. If the E-field at the location of the atom is not maximized, the cooperativity is accordingly reduced from its maximum value. The reduction factor in this case can be calculated using the ratio of the E-field at the atom's location to its peak value: $|E(r_{\text{atom}})/E(\text{peak})|^2$.

$$\frac{P_{4\pi}}{P_{\text{in}}} = \frac{2\eta}{(1 + \eta)^2}$$

$$\frac{P_{\text{tr}}}{P_{\text{in}}} = \frac{1}{(1 + \eta)^2}$$

On the atomic and cavity resonance, the ratio of the scattered power, $P_{4\pi}$, and the transmitted power, P_{tr}, to the incident power,

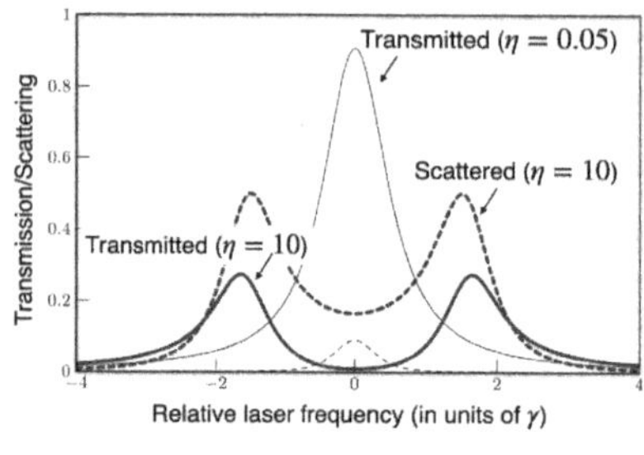

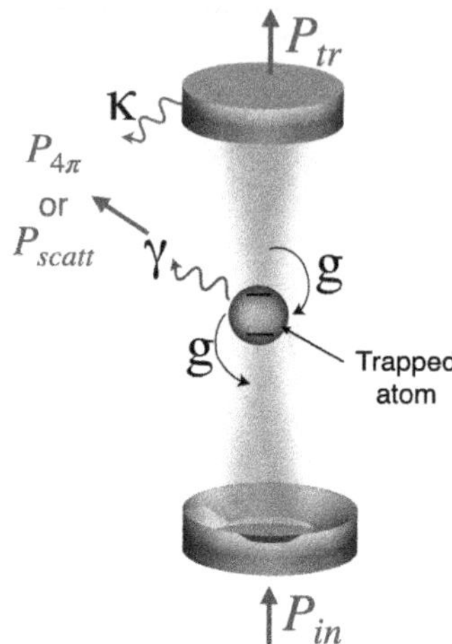

P_{in}, can be expressed in terms of the cooperativity. It can be seen that at high cooperativities, the transmission and scattering are suppressed to the point where most light is reflected from the cavity. In other words, a single atom resonantly interacting with a cavity photon can induce such a strong absorption that it effectively shifts the cavity resonance condition (by inducing a large refractive index, effectively changing the cavity length).

In the case of total cavity blocking on the resonance condition, the incident photon is reflected from the cavity with a high probability, with little chance of scattering or loss. This however does not indicate that the photon will not enter the cavity or not interact with the atom. It is the interaction with the atom that gives rise to cavity blocking.

2.6.3. *Light–Atom Entanglement Inside a Cavity*

Consider an optical state in a superposition of 0 or 1 photon state that is incident on a cavity containing a single atom. In the high cooperativity regime and on atomic and cavity resonance, the atom can block the cavity only if it is prepared in the ground energy state $|g\rangle$. If the atom is also prepared in a superposition state of two energy levels $|g\rangle$ and $|e\rangle$, the outcome of the interaction is an entangled state

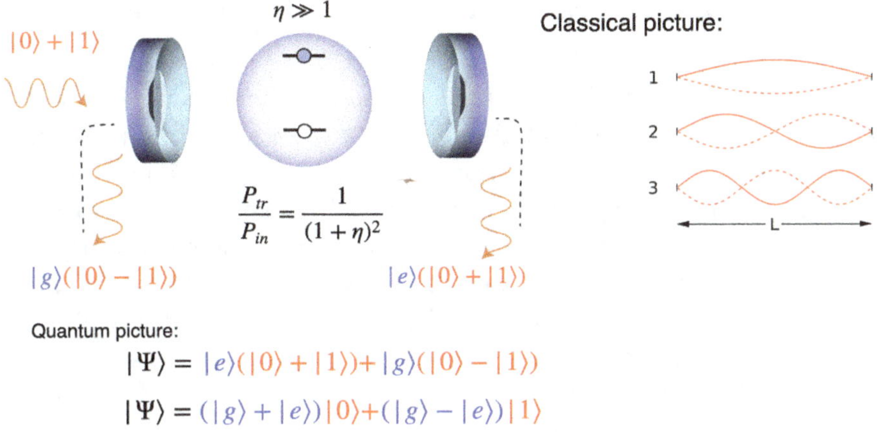

$$|\Psi\rangle = |e\rangle(|0\rangle + |1\rangle) + |g\rangle(|0\rangle - |1\rangle)$$

$$|\Psi\rangle = (|g\rangle + |e\rangle)|0\rangle + (|g\rangle - |e\rangle)|1\rangle$$

of the photon and the atom. When the atom is in the ground state, the optical state is fully reflected, (cavity blocking) and as a result, the phase of the reflected light changes by 180 degrees. This π phase change is characteristic of the reflection from a surface with a higher refractive index (illustrated by the classical wave picture on the top right for the first three fundamental modes). In optical cavities, the π phase change is needed to satisfy the interference condition at the input and the output, and achieve the cavity resonance condition.

Therefore, when the atom is in the $|g\rangle$ state, the input optical state changes from $|0\rangle + |1\rangle$ to $|0\rangle - |1\rangle$, ignoring the normalization coefficient. On the other hand, when the atom is in the $|e\rangle$ state, the optical state is transmitted unaffected. The total state of the system can no longer be described by the atom and field states separately. Now, it should be written as a coherent superposition of the two outcomes.

We see that in this strong coupling regime, simply impinging an optical state to the atom–cavity system results in the deterministic (not probabilistic) generation of atom–photon entanglement. An interaction of this kind can also be used to create photon–photon entanglement by sending a subsequent photon into the cavity.

In the simple reflection picture, we see that interaction can be interpreted as the atom or photon experiencing a 180-degree phase shift since the global phase does not matter.

2.7. Two-photon Transitions in Three-Level Atoms

So far, we have considered the simple picture of a photon resonantly interacting with a two-level atom. In reality, the atom has multiple energy levels and the light frequency may not match a particular atomic frequency. In this section, we consider a more general case of light–atom interactions, where two light fields interact with an atom with three-energy levels. This interaction scheme is important in quantum communication and it is frequently used to coherently manipulate, store, and control optical and atomic qubits.

2.7.1. *Three-Level Atom*

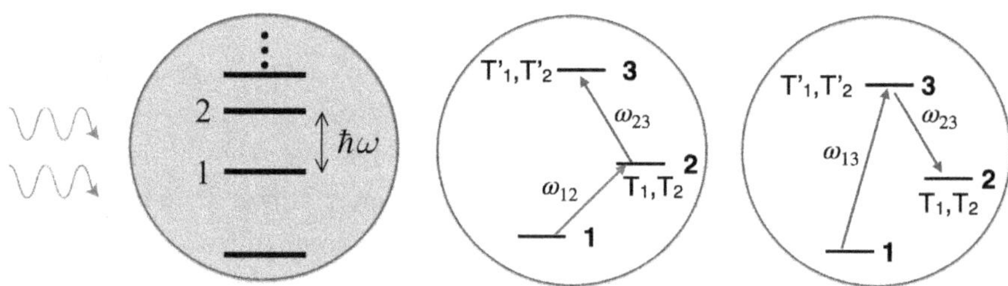

The energy levels of atoms are determined by the electron configuration and interactions with internal and external fields. The electrons inside an atom can reside in different energy levels and the spacing between adjacent energy levels can vary and may not be equal. As a result, an atom can interact with a single or multiple optical field via separate transitions. Which transitions are excited depends on the relative frequency of the light to the energy spacing of the atom, light polarization, selection rules, and more.

The middle figure above shows one way that two electric fields with frequencies ω_{12} and ω_{23} can interact with two transitions of a three-level atom. In this cascaded transition, if the atom is initially prepared in the energy level $|1\rangle$, the two fields can derive the atom from $|1\rangle$ to $|3\rangle$ through level $|2\rangle$. This is true if the population decay rate of level $|2\rangle$, $1/T_1$, is slow compared to the Rabi frequency of the two fields. The coherence times T_2 and T_2' determine whether coherent mapping from the field to the atom can be achieved within a given timescale.

Another example of a two-photon transition is shown on the right, known as Λ-transition, where the two fields with frequencies ω_{13} and ω_{23} excite the atom from $|1\rangle$ to $|2\rangle$ via the excited state $|3\rangle$.

Such two-photon transitions are frequently used in atomic spectroscopy. In some cases, a particular transition may be forbidden by the selection rules, and as such, two-photon excitations enable us to indirectly excite the atom. The addition of an EM field can also be used as a coherent control knob to control the coupling of the other field to and from an atom.

2.7.2. *Resonant Interaction*

Electromagnetically Induced Transparency

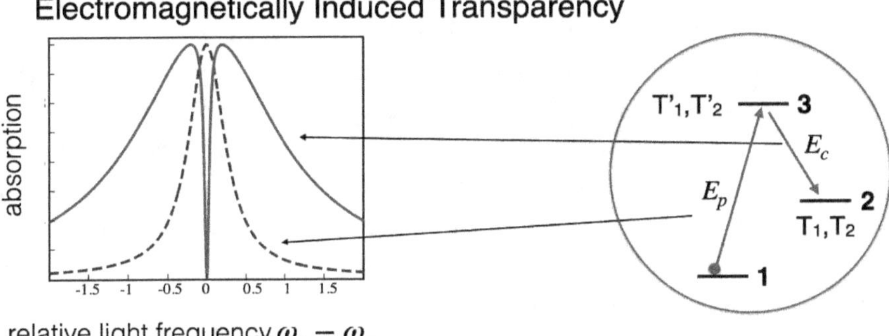

Consider the Λ-transition shown above. The interaction Hamiltonian of the system can be written as

$$H_{\text{int}} = -\hbar/2(\Omega_p \hat{\sigma}_{31} e^{i\Delta_p t} + \Omega_c \hat{\sigma}_{32} e^{i\Delta_c t})$$

where Ω_p and Ω_c denote the Rabi frequency associated with the probe field, E_{p}, and the control field, E_{c}, respectively. The atomic operators are written as $\sigma_{ij} = |i\rangle\langle j|$ and Δ_p and Δ_c which are the frequency difference of the probe and control fields from 1–3 and 1–2 transitions, respectively. We assume $\Delta_p = \Delta_c = \Delta$.

Under resonant interactions, and when the control field is much stronger than the probe field $E_p \ll E_c$, the two absorption paths can interfere, resulting in suppression of probe absorption. This regime is known as electromagnetically induced transparency (EIT) (Harris *et al.*, 1990; Fleischhauer *et al.*, 2005), where the two possible pathways in which light can be absorbed by an atom undergo quantum interference. Due to the destructive interference, the absorption vanishes, a narrow transparency window is opened inside the opaque medium, and probe propagation is accompanied by strong dispersion.

In the absence of a strong control field, the probe light is maximally absorbed on atomic resonance, as seen by the dashed curve (plot of susceptibility of the medium) on the left. As the frequency of the probe field changes away from the atomic resonance, the absorption strength decreases. If we now introduce a strong control field, the interference between the two fields at the resonance condition opens a transparency window into the medium, and the probe is now transmitted, meaning it is no longer absorbed by the atom. The probe propagation through the EIT window is accompanied by strong dispersion given by the steep slope of the susceptibility curve near the resonance.

As a result, the probe propagation occurs with low group velocity, enabling us to delay it or even store its information in the form of atomic coherence between energy levels 1 and 2. It can be shown that the group velocity of the weak probe field is proportional to the intensity of the control field.

In the limit that the control field is much stronger than the other:

- absorption vanishes,
- group velocity of weak probe field: $v_g \propto E_c^2$.

2.7.3. *Off-Resonant Raman Transition*

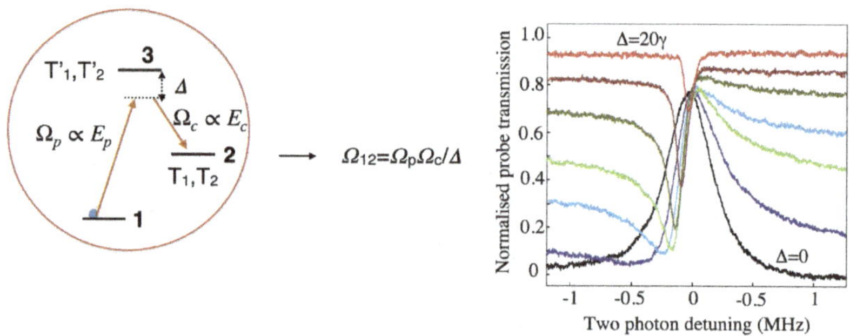

When the probe field is pulsed and the pulse is compressed (due to reduced group velocity) to fully fit inside the atomic medium, reducing the control intensity to zero can practically "stop the light". At this point, it is not the light that is stopped, but it is its information that is mapped to the atomic coherence, $|1\rangle\langle 2|$. Subsequently, increasing the control field power after time τ can result in the retrieval of the probe pulse, given the atomic coherence time is longer than τ. This method can be used to store quantum optical information. This has applications in quantum communication and computation.

Off-resonant interactions like the one shown above on the left of the preceding figure are also frequently used to coherently control atomic coherence. An off-resonant light will not be absorbed by an atom. However, when both the control and probe fields are detuned from the excited state while resonantly interacting with levels 1 and 2, the two transmission paths can interfere and result in absorption. The effective coupling strength between levels 1 and 2 is then given by $\Omega_{12} = \Omega_p\Omega_c/\Delta$, assuming $\Delta \gg \gamma$ where γ is the excited state linewidth. At large detunings and strong control powers, the probe field can be coherently absorbed by the two lower energy states of the atom.

Off-resonant interaction of two fields can give rise to absorption (two-photon resonance) effectively creating a two-level system.

The experimental data on the right shows the probe transmission through rubidium (Rb) atoms in the presence of a strong control field for different detunings, Δ. As expected, close to resonance ($\Delta \sim 0$), the probe is transmitted through the EIT window. As the detuning increases, the transmission changes to absorption.

This off-resonant coupling is called the Raman transition and it can also be used for the storage of a probe pulse by pulsing on and off a control laser. The process is also routinely used to pump (or prepare) atoms in either of the lower energy states when the direct transition is not allowed or effective.

An effective Rabi frequency between the two lower energy states can be engineered and the duration of the laser pulses can be chosen to precisely control the population of the atom or to create arbitrary superposition states.

We can try to obtain a more rigorous description of the interaction in this Λ-configuration by writing the Schrödinger equation. Assuming a generic atomic state, $|\psi\rangle = C_1|1\rangle + C_2|2\rangle + C_3|3\rangle$, we can derive equations of motions of probabilities C_i. We can define the Rabi frequency of the two fields as $\Omega_{p,c} = \frac{\langle 3|\hat{d}.E_{p,c}|1,2\rangle}{\hbar}$. Writing the Schrödinger equation, we then find

$$\dot{C}_1 = -(i\delta_{\text{ac},p} + \gamma_p/2)C_1 - i\frac{\Omega_{\text{Ram}}}{2}C_2$$

$$\dot{C}_2 = -(i\delta_{\text{ac},c} + \gamma_c/2)C_2 - i\frac{\Omega_{\text{Ram}}}{2}C_1 \qquad (2.1)$$

$$\dot{C}_3 = 0$$

Above, we assume that C_3 dynamics is much slower than timescales associated with Δ and Γ and ignore its time dependency. We have defined the ac-Stark frequency shift as $\delta_{\text{ac},p/c} = \frac{\Omega_{p/c}^2}{4\Delta}$, Raman scattering rate as $\gamma_{p,c} = \frac{\Omega_{p,c}^2}{4\Delta^2}\gamma$, effective Raman coupling rate as $\Omega_{\text{Ram}} = \frac{\Omega_1\Omega_2}{2\Delta}$, and spontaneous emission rate of level 3, γ. The time derivative of the reduced density matrix element corresponding to coherence is then given by

$$\dot{\rho}_{21} = C_2^*\dot{C}_1 + \dot{C}_2^*C_1 = -\left(i\delta_{\text{Ram}} + \frac{\gamma_{12}}{2}\right)\rho_{21} - i\frac{\Omega_{\text{Ram}}}{2}(\rho_{22} - \rho_{11})$$

where Raman detuning is defined as $\delta_{\text{Ram}} = \delta_{\text{ac},p} - \delta_{\text{ac},c}$ and total decay rate assuming an additional ground-state decoherence rate, γ_0, is $\gamma_{12} = \gamma_p + \gamma_c + \gamma_0$. The equation suggests coherence between the two lower energy states driving by Ω_{Ram}. This shows how information from an EM field is mapped to the atomic coherence.

2.8. Dynamics of Density Operator

In this section, we provide more detailed explanation about density matrix and derivation of light-atom interaction dynamics (equations of motion) using the density matrix formalism. We have introduced the density matrix as another representation for the quantum state wave function. In most practical cases, the state of a system is a mixed state consisting of different wave functions. If the system is described by wave function $|\psi_n\rangle$ with probabilities p_n, then the appropriate density operator is

$$\hat{\rho}(t) = \sum_n p_n |\psi_n(t)\rangle \langle \psi_n(t)|.$$

Here $\hat{\rho}$ is the density operator and for any complete set of basis states it can be represented by a matrix (the density matrix). We can write $|\psi_n\rangle$ above in terms of its basis states $\{|i\rangle\}$, and arrive at

$$\hat{\rho} = \sum_{ij} |i\rangle \langle i|\hat{\rho}|j\rangle \langle j| = \sum_{ij} |i\rangle \rho_{ij} \langle j|.$$

The diagonal matrix element ρ_{ii} are population terms or the probabilities of finding the system in state $|i\rangle$; the off-diagonal elements ρ_{ij} are coherences between states i and j. A non-zero off-diagonal element suggests phase coherence (e.g. coherent suppression) between different states and thus is a measure of "quantumness". The density operator is hermitian, i.e. $\hat{\rho}^\dagger = \hat{\rho}$. The hermitian operators have real eigenvalues that correspond to measurable quantities. In Eq. (2.2), if states $\{|\psi_n\rangle\}$ are themselves orthonormal basis functions, the eigenvalues of the density operators are simply probabilities p_n that are bound between 0 and 1 and $\sum p_n = 1$ as seen below:

$$\mathrm{Tr}[\hat{\rho}] = \sum_i \langle i|\hat{\rho}|i\rangle = \sum_n p_n \sum_i |\langle i|\psi_n\rangle|^2 = \sum_n p_n = 1$$

A pure quantum state is represented as a linear superposition of basis states, $|i\rangle$ as

$$|\psi\rangle = \sum_i \lambda_i |i\rangle$$

Some properties of pure states include:

- $\mathrm{Tr}[\rho] = \sum_i \lambda_i$
- $\mathrm{Tr}[\rho] = 1$
- $\rho^2 = \rho$

In reality, there is no pure state and all states measured in a lab are a statistical mixture of pure states, i.e. mixed states as described by Eq. (2.2). In the case of a mixed state

$$Tr[\rho^2] < 1.$$

Note that in Eq. (2.2), each wave function $|\psi_n(t)\rangle$ is written in terms of the basis function (e.g. $|i\rangle$) and when multiplied by $\langle\psi_n(t)|$ the product can not be written as a pure state density operator, i.e.

$$\hat{\rho} = \sum_{ij} \rho_{ij}|i\rangle\langle j|. \tag{2.2}$$

The wave function of a composed system, e.g. atom-light system, can be written as a tensor product $|\psi_{A_i}\rangle \times |\psi_{L_j}\rangle$. The corresponding density matrix of the combined system ρ can be written as

$$\hat{\rho} = \sum_{ij}\sum_{kl} \rho_{ij}^{kl}|ij\rangle\langle kl|.$$

It is sometimes useful to consider the elements of a system separately. In the case of light interacting with atoms, we might be interested to describe the evolution of the light field as the result of interaction with the atoms or vice versa. In such case, we can define the reduced density matrix for system A only as

$$\hat{\rho}_A = \sum_{i}\sum_{k} \sigma_{ik}|i\rangle\langle k|$$

where σ_{ik} is the partial trace over system L, $\sigma_{ik} = \sum_j \rho_{ij}^{kj}$.

Alternatively, if $\hat{\rho}$ is known, the expectation value of any operator $\hat{O}$ can be found as:

$$\langle\hat{O}\rangle = \sum_{n} p_n\langle\psi_n|\hat{O}|\psi_n\rangle$$

$$= \sum_{ij}\sum_{n} p_n\langle\psi_n|i\rangle\langle i|\hat{O}|j\rangle\langle j|\psi_n\rangle$$

$$= \sum_{ij} O_{ij}\rho_{ji} = \mathrm{Tr}[\hat{O}\hat{\rho}].$$

We will use this form when calculating the expectation value of atomic operators from density matrix elements.

Taking the time-derivative of the density operator we can arrive at the Schrödinger — Von Neumann equation

$$\frac{\partial \hat{\rho}}{\partial t} = \sum_n p_n (\partial_t |\Psi_n(t)\rangle)\langle\Psi_n(t)| + |\Psi_n(t)\rangle(\partial_t \langle\Psi_n(t)|)$$

$$= \frac{1}{i\hbar}\sum_n p_n \hat{H}|\Psi_n(t)\rangle\langle\Psi_n(t)| - \frac{1}{i\hbar}\sum_n p_n|\Psi_n(t)\rangle\langle\Psi_n(t)|\hat{H}$$

$$= \frac{1}{i\hbar}[\hat{H}, \hat{\rho}]$$

where $\hat{H}$ is the Hamiltonian of the system. Note that this equation is derived in the Schrödinger representation, where the wave functions are considered time-dependent but the operators are not.

Going to the Heisenberg representation, where the operators are considered to be time-dependent, we can describe the operator dynamics by the following equations (note the difference between the derivative and partial derivative with respect to time):

$$\frac{d\hat{O}}{dt} = \frac{\partial \hat{O}}{\partial t} + \frac{1}{i\hbar}[\hat{O}, \hat{H}].$$

This equation can be used to derive equations of motion describing the dynamics of operators or their expectations values. When there is no explicit time dependency for the operators, i.e. $d\hat{O}/dt = 0$, the equation will look similar to the Von Neumann equation. We use this equation later to derive equations of motion for expectation values of the atomic operators.

Since the Hamiltonian is hermitian, there are no imaginary terms in the preceding equation, suggesting no decay. However, the decay of population or decoherence needs to be considered for any realistic processes. We can add those terms manually to the equation above to account for various decays.

The formal way of dealing with decoherence is to follow the Lindblad master equation where additional terms are added to the above equation modeling decay (e.g. spontaneous emission) and decoherence (e.g. due to magnetic field fluctuations) processes in the system.

Consider the state of a single atom initially ($t = 0$) in the excited state given by $|\psi_0\rangle$. Atom undergoes spontaneous emission after time δt with probability P forcing the atom to decay to the ground state $|\psi_g\rangle$. With probability $1 - P$, atom remains in the excited state after time δt whose state is described by $|\psi_{\delta t}\rangle$. The state of the system after time δt is a mixed state (see Ref. [Gerry and Knight (2005)] for more details)

$$\hat{\rho} = |\psi_{\delta t}\rangle\langle\psi_{\delta t}| = (1 - P)|\psi_e\rangle\langle\psi_e| + P|\psi_g\rangle\langle\psi_g|.$$

where the probability P can be written in terms of the decay rate, Γ, and field's ladder operators as

$$P = (\Gamma\delta t)\langle\psi_0|\hat{a}^\dagger\hat{a}|\psi_0\rangle.$$

If atoms decay, the state of the atom "jumps" to

$$|\psi_g\rangle = \frac{\hat{a}|\psi_0\rangle}{\langle\psi_0|\hat{a}^\dagger\hat{a}|\psi_0\rangle^{1/2}}.$$

Otherwise, the state after time δt is found by simple time evolution and normalization

$$|\psi_e\rangle = \frac{e^{-i\delta t\hat{H}}|\psi_0\rangle}{\langle\psi_0|e^{i\delta t\hat{H}^\dagger}e^{-i\delta t\hat{H}}|\psi_0\rangle^{1/2}}.$$

The mixed-density operator is then given by

$$|\psi_{\delta t}\rangle\langle\psi_{\delta t}| = (\Gamma\delta t)\ \hat{a}|\psi_0\rangle\langle\psi_0|\hat{a}^\dagger$$
$$+ \left(1 - \frac{i}{\hbar}\hat{H}\delta t - \frac{\Gamma}{2}\delta t\hat{a}^\dagger\hat{a}\right)|\psi_0\rangle\langle\psi_0|\left(1 + \frac{i}{\hbar}\hat{H}\delta t - \frac{\Gamma}{2}\delta t\hat{a}^\dagger\hat{a}\right)$$
$$\simeq |\psi_0\rangle\langle\psi_0| - \frac{i}{\hbar}\delta t[\hat{H}, |\psi_0\rangle\langle\psi_0|]$$
$$+ \frac{\Gamma}{2}\delta t\{2\hat{a}|\psi_0\rangle\langle\psi_0|\hat{a}^\dagger - \hat{a}^\dagger\hat{a}|\psi_0\rangle\langle\psi_0| - |\psi_0\rangle\langle\psi_0|\hat{a}^\dagger\hat{a}\}.$$

Averaging over all states and in the limit of $\delta t \to 0$, we obtain

$$\frac{\partial\rho}{\partial t} = -\frac{i}{\hbar}[\hat{H}, \rho] + \frac{\Gamma}{2}\{2\hat{a}\rho\hat{a}^\dagger - \hat{a}^\dagger\hat{a}\rho - \rho\hat{a}^\dagger\hat{a}\}.$$

This equation is known as the master equation. In the case where more decay processes occur, the last term should be the sum of all decay processes with corresponding jump operators, e.g. $\hat{a}$ and $\hat{a}^\dagger$.

2.9. Interaction of Light with Atoms in Free Space

Consider a two-level atom with ground, $|g\rangle$, and excited, $|e\rangle$, state energy levels at frequency spacing ω_{eg} interacting with EM field of frequency ω_L. We have previously defined the atomic operators in terms of the Pauli spin matrices, $\hat{\sigma}_x$, $\hat{\sigma}_y$ and $\hat{\sigma}_z$ that represents the angular momentum of the electron spin, as

$$\hat{\sigma}^\dagger = \hat{\sigma}_{eg} = \hat{\sigma}_x + i\hat{\sigma}_y = |e\rangle\langle g|$$

$$\hat{\sigma} = \hat{\sigma}_{ge} = \hat{\sigma}_x - i\hat{\sigma}_y = |g\rangle\langle e|$$

$$\hat{\sigma}_z = |e\rangle\langle e| - |g\rangle\langle g|.$$

The operators $\hat{\sigma}^\dagger$ and $\hat{\sigma}$ are atomic ladder operators and account for atomic coherence and $\hat{\sigma}_z$ represents the population difference of atom in the two states.

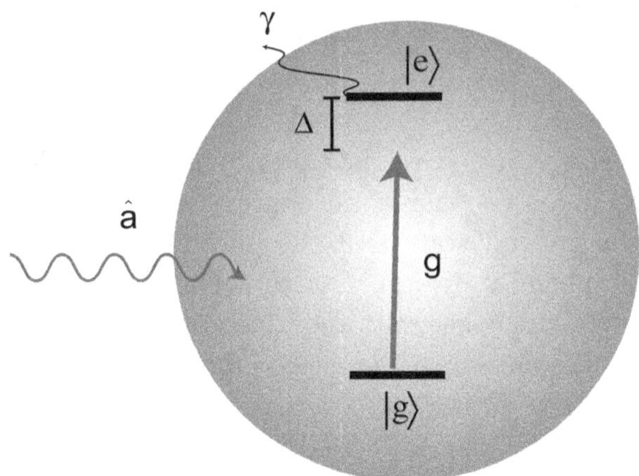

A two-level atom, with energy levels $|g\rangle$, and $|e\rangle$ and frequency spacings ω_{eg}, interacts an optical field, $\hat{a}$ with vacuum Rabi frequency (coupling rate) g.

2.9.1. *Schrödinger Picture*

We first consider and describe the interaction of two-level atoms, with energy levels $|g\rangle$, and $|e\rangle$ and frequency spacings ω_{eg}, with an optical field, $\hat{a}$ of frequency ω_L and vacuum Rabi frequency (coupling rate) g, in the Schrödinger picture (see the figure below).

The total Hamiltonian of the system can be written as

$$\hat{H}/\hbar = (1/2 + \hat{a}^\dagger \hat{a})\omega_L + \omega_{eg}\hat{\sigma}^\dagger\hat{\sigma} + g(\hat{\sigma}\hat{a}^\dagger + \hat{\sigma}^\dagger\hat{a}).$$

The first term above is the light Hamiltonian, the second term is the atom Hamiltonian and the third term is the light-atom interaction Hamiltonian. The interaction term has two other components, $\hat{\sigma}^\dagger\hat{a}^\dagger$ and $\hat{\sigma}\hat{a}$ that oscillate at frequency $\pm(\omega_{eg} + \omega_L)$ that are much larger than the frequency of the other two terms $\pm(\omega_{eg} - \omega_L)$ and thus quickly average to zero. We call this approximation, the rotating wave approximation (RWA). We can also drop the term corresponding to the constant energy $1/2\hbar\omega_L$ by changing the energy reference.

A generic solution of wave function to the Schrödinger equation

$$i\hbar\frac{\partial}{\partial t}|\psi\rangle = \hat{H}|\psi\rangle$$

can be written as

$$|\psi\rangle = c_{g,1}|g\rangle|1\rangle + c_{e,0}|e\rangle|0\rangle$$

where coefficients $c_{g/e,1/0}$ are the probability amplitudes and $|g\rangle|1\rangle$, for example, indicates the total state of the system when the atom is in the ground state and the photon is not absorbed. The probability amplitudes of other states such as $|g\rangle|0\rangle$ and $|e\rangle|1\rangle$ are zero, considering the conservation of energy.

Plugging this solution to the Schrödinger equation we have

$$\dot{c}_{g,1}|g\rangle|1\rangle + \dot{c}_{e,0}|e\rangle|0\rangle = -i\omega_L\hat{a}^\dagger\hat{a}|\psi\rangle$$
$$- i\omega_{eg}\hat{\sigma}^\dagger\hat{\sigma}|\psi\rangle$$
$$- ig(\hat{\sigma}\hat{a}^\dagger + \hat{\sigma}^\dagger\hat{a})|\psi\rangle.$$

Considering that $\hat{a}$ and $\hat{a}^\dagger$ only operate on the optical states and $\hat{\sigma}^\dagger$ and $\hat{\sigma}$ only on the atomic states, we have the following operations:

$$\hat{a}^\dagger\hat{a}|g\rangle|1\rangle = |g\rangle|1\rangle$$

$$\hat{a}^\dagger\hat{a}|e\rangle|0\rangle = 0$$

$$\hat{\sigma}^\dagger\hat{\sigma}|g\rangle|1\rangle = 0$$

$$\hat{\sigma}^\dagger\hat{\sigma}|e\rangle|0\rangle = |e\rangle|0\rangle$$

$$\hat{\sigma}\hat{a}^\dagger|g\rangle|1\rangle = 0$$

$$\hat{\sigma}\hat{a}^\dagger|e\rangle|0\rangle = |g\rangle|1\rangle$$

$$\hat{\sigma}^\dagger\hat{a}|g\rangle|1\rangle = |e\rangle|0\rangle$$

$$\hat{\sigma}^\dagger\hat{a}|e\rangle|0\rangle = 0$$

Note that $\hat{a}^\dagger|1\rangle = 0$ and $\hat{\sigma}^\dagger|e\rangle = 0$ because we only have one photon and one excited state. Using the above operations, we can rewrite the Schrödinger equation as

$$\dot{c}_{g,1}|g\rangle|1\rangle + \dot{c}_{e,0}|e\rangle|0\rangle = -i\omega_L c_{g,1}|g\rangle|1\rangle$$
$$- i\omega_{eg}c_{e,0}|e\rangle|0\rangle$$
$$- ig(c_{g,1}|e\rangle|0\rangle + c_{e,0}|g\rangle|1\rangle).$$

Equating the coefficients of $|e\rangle|0\rangle$ and $|g\rangle|1\rangle$ terms on both sides we arrive at following dynamics:

$$\dot{c}_{g,1} = -i\omega_L c_{g,1} - igc_{e,0}$$

$$\dot{c}_{e,0} = -i\omega_{eg}c_{e,0} - igc_{g,1}.$$

To simplify further, we rewrite the above equations in a frame rotating at the optical frequency, i.e.

$$c_{g,1} \to c_{g,1}e^{-i\omega_L t}$$

$$c_{e,0} \to c_{e,0}e^{-i\omega_L t}.$$

We then find the new equations defining $\Delta = \omega_{eg} - \omega_L$, as

$$\dot{c}_{g,1} = -igc_{e,0}$$

$$\dot{c}_{e,0} = -i\Delta c_{e,0} - igc_{g,1}.$$

We can add a phenomenological decay rate γ for damping of atom's coherence between the two levels to obtain

$$\dot{c}_{g,1} = -igc_{e,0}$$

$$\dot{c}_{e,0} = -(\gamma + i\Delta)c_{e,0} - igc_{g,1}.$$

Taking time derivative of the second equation and replace $\dot{c}_{g,1}$ from the first equation we have

$$\ddot{c}_{e,0} + (\gamma + i\Delta)\dot{c}_{e,0} + g^2 c_{g,1} = 0.$$

The equation above indicates damped oscillation with the solution given by

$$c_{e,0} = c_0 e^{-(\gamma + i\Delta)t} \cos\frac{\Omega t}{2}.$$

where effective oscillation frequency, known as Rabi frequency, $\Omega = \sqrt{\Delta^2 + \gamma^2 + 4g^2}$. We see that the probability amplitude of the atom in the excited state oscillates at frequency $\Omega/2$, which in the ideal case when $\gamma = \Delta = 0$, it becomes g. This oscillation is known as Rabi oscillation and g is referred to as vacuum Rabi frequency. In presence of damping the oscillation vanishes in a time scale of $1/\gamma$.

2.9.2. *Heisenberg Picture*

We now use the same the Heisenberg picture to write the equations of motion for the expectation values of atomic operators for the case of light interacting with two- and three-level atoms. For a two-level atom shown in the preceding figure we can write the following set of equations for expectation values using the Heisenberg equation,

$$\partial_t \sigma_{ee} = i\Omega/2(\sigma_{eg} - \sigma_{ge}) - \Gamma\sigma_{ee}$$

$$\partial_t \sigma_{gg} = -i\Omega/2(\sigma_{eg} - \sigma_{ge}) + \Gamma\sigma_{ee}$$

$$\partial_t \sigma_{ge} = -(\gamma + i\Delta)\sigma_{ge} - i\Omega/2(\sigma_{ee} - \sigma_{gg})$$

where $\Delta = \omega_{eg} - \omega$ is the frequency difference (detuning) of the EM field with respect to the atomic transition, Ω is the Rabi frequency of the optical field, and Γ is the excited state population decay rate. In the equations above we have included a phenomenological decay rate to account for loss and decoherence. Assuming an additional damping rate γ_0 that accounts for the dephasing processes, for example atom-atom collisions or magnetic field fluctuations, the total decoherence rate γ can be written as

$$\gamma = \Gamma/2 + \gamma_0.$$

Note that to arrive at the above equations of motions for the atomic population and coherence we substituted $\langle \hat{\sigma}_{ij} \rangle = \sum_{ij} \sigma_{ij}\rho_{ji} = \text{Tr}[\hat{\sigma}\hat{\rho}]$ by σ_{ij}.

The steady-state solution of the above equations can be found in the limit that the dynamics of population and decoherence are slow compared to other time scales of the system, e.g. Γ, i.e. $\partial_t \sigma_{ij} = 0$. In this limit, solving for the population term we find

$$\sigma_{ee}(t \to \infty) = \frac{\Gamma}{4\gamma} \frac{s}{(1+s)^2}$$

where s is the saturation parameter and is defined as

$$s = \frac{\Omega^2/\gamma\Gamma}{1+\Delta^2/\gamma^2}.$$

In the resonant condition, $\Delta = 0$, we have

$$\sigma_{ee}(t \to \infty, \Delta = 0) = \frac{\Omega^2/4\gamma^2}{(1+\Omega^2/\gamma\Gamma)^2}.$$

Now consider the interaction of a three-level atom with two light fields (probe and control fields) as shown in the figure below. The interaction Hamiltonian of this system can be written as

$$H_{\text{int}} = -\hbar/2(\Omega_p \hat{\sigma}_{31} e^{i\Delta_p t} + \Omega_c \hat{\sigma}_{32} e^{i\Delta_c t} + H.c.)$$

where γ_{ij} denotes the decay rates between levels i and j, $\Omega_p = gE_p$ and Ω_c denote the Rabi frequencies of the probe and control fields, respectively, g is probe-light-atom coupling rate, and E_p or $\hat{E}_p$, which is treated as quantum field, is the probe's electric field amplitude that is much weaker than the control field (treated as classical field). The frequency detunings of the probe and control fields from the excited state are Δ_p and Δ_c, respectively.

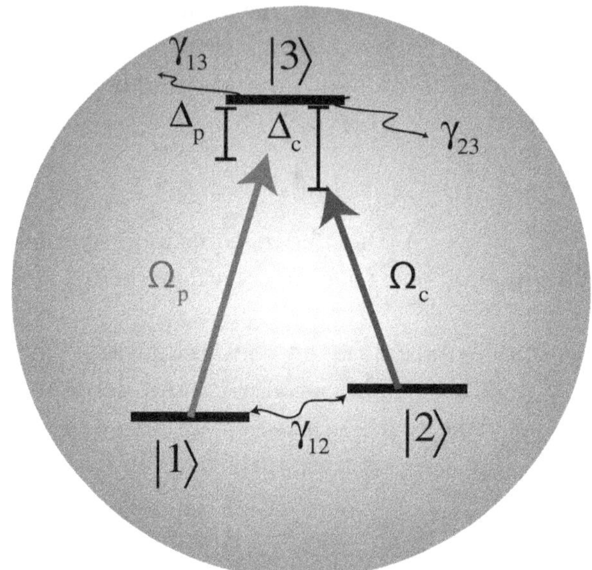

A three-level atom, with energy levels $|1\rangle$, $|2\rangle$, $|3\rangle$ and frequency spacings ω_{31} and ω_{32}, interacts two fields with Rabi frequency Ω_c and Ω_p.

Similar to the case of the two-level atom, we can arrive at the equations of motion for the atomic operators' expectation values as

$$\dot{\sigma}_{12} = -(\gamma_{12} - i\delta)\sigma_{12} + i\Omega_c^{\star}/2\sigma_{13} - i\Omega_p\sigma_{32}$$

$$\dot{\sigma}_{13} = -(\gamma_{13} - i\Delta)\sigma_{13} + i\Omega_p/2(\sigma_{11} - \sigma_{33})$$

$$\dot{\sigma}_{32} = -(\gamma_{23} - i\Delta)\sigma_{32} + i\Omega_c^{\star}/2(\sigma_{33} - \sigma_{22}) - i\Omega_p^{\star}\sigma_{12}$$

where * represents complex conjugate operation, and $\delta = \Delta_p - \Delta_c = \omega_p - \omega_c - \omega_{12}$ is the two-photon detuning and $\Delta = \Delta_p = \omega_{13} - \omega_p$ is the one-photon detuning. Below we take $\gamma_{13} = \gamma_{23} = \gamma$.

We can solve for σ_{13} in steady state as

$$\sigma_{13} = \frac{4\delta(|\Omega_c|^2 - 2\delta\Delta) - 4\Delta\gamma_{12}^2}{||\Omega_c|^2 + (\gamma_{13} + i2\Delta)(\gamma_{12} + i2\delta)|^2}$$
$$+ i\frac{8\delta^2\gamma_{13} + 2\gamma_{12}(|\Omega_c|^2 + \gamma_{12}\gamma_{13})}{||\Omega_c|^2 + (\gamma_{13} + i2\Delta)(\gamma_{12} + i2\delta)|^2}$$

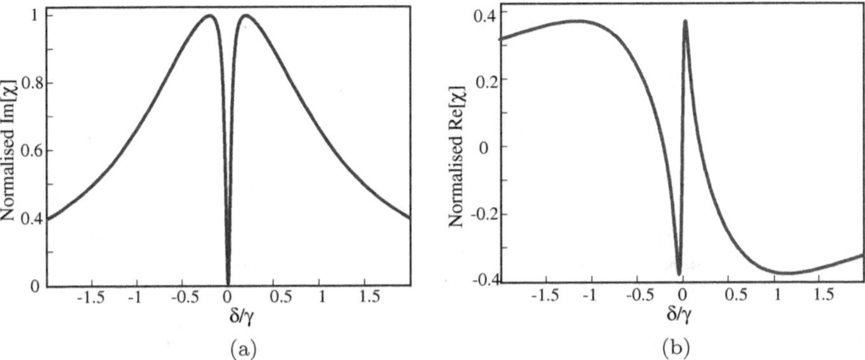

Normalized imaginary (a) and real (b) parts of susceptibility plotted as a function of two-photon detuning, δ at zero one-photon detuning $\Delta = 0$.

that is proportional to the susceptibility seen by the probe light:

$$\chi^{(1)}(\omega_p) = \mathcal{G}\sigma_{13}$$

where $\mathcal{G} = \mu_{13}^2/\epsilon_0\hbar V$, μ_{13} is the dipole moment of the transition, and V is the interaction mode volume. The real and imaginary parts of the susceptibility are plotted in the following figure as a function of two-photon detuning, δ, for $\Delta = 0$.

The transmission and refractive index of the probe field can be expressed using the real and imaginary parts of $\chi^{(1)}$

$$T(\omega) = e^{-\text{Im}[\chi(\omega)]k_p L},$$

$$n(\omega) = 1 + \text{Re}[\chi(\omega)]/2.$$

where L is the interaction length and $k_p = 2\pi/\lambda_p$ is the wavenumber of the light with wavelength λ_p. The transmission of a probe light interaction with three-level atoms in the EIT condition is plotted for two different optical densities (d) and control Rabi frequencies (Ω)

As seen in part (a) of the above figure, at zero one- and two-photon detunings, the absorption vanishes. At low probe and control powers, this is fundamentally explained by quantum interferences between the two fields. This regime is known as electromagnetically induced transparency (EIT). At high powers, the physics changes

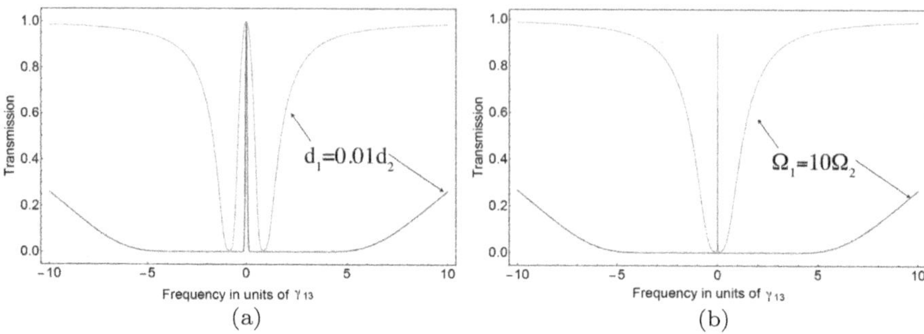

Transmission of probe light through an ensemble of three-level atoms in the EIT condition plotted for two different optical densities (d) and control Rabi frequencies (Ω).

and the transmission occurs due to the shift in the atomic transition caused by the strong light field (i.e. light shift or ac Stark shift). The steep variation in the real part of the susceptibility gives rise to the reduced group velocity of the probe light

$$v_g = \frac{d\omega_p}{dk_p} = \frac{c}{n + \omega_p \partial n / \partial \omega_p} \simeq \frac{\Omega_c^2 L}{d\gamma}$$

where $d = g^2 L / \gamma c$ is defined as the optical depth of the medium. In the EIT regime, the probe light's group velocity can be reduced by lowering the power of the control field or increasing the interaction strength. Experimental group velocities as low as 17 m/s have been demonstrated. The transmission bandwidth however also scales with the control field power. So to achieve large delay-bandwidth products one must increase the optical depth.

In the case where light interacts with N identical atoms, one can show that by replacing g with $g\sqrt{N}$, the equations derived above can be used to describe light interaction with the ensemble of atoms [Gorshkov *et al.* (2007)], where L is the length scale of interaction or size of the ensemble.

2.9.3. *Four-Wave Mixing in Four-Level Systems*

Interaction of light with more energy levels of atoms can result in more complex processes. Here we discuss the interaction of three

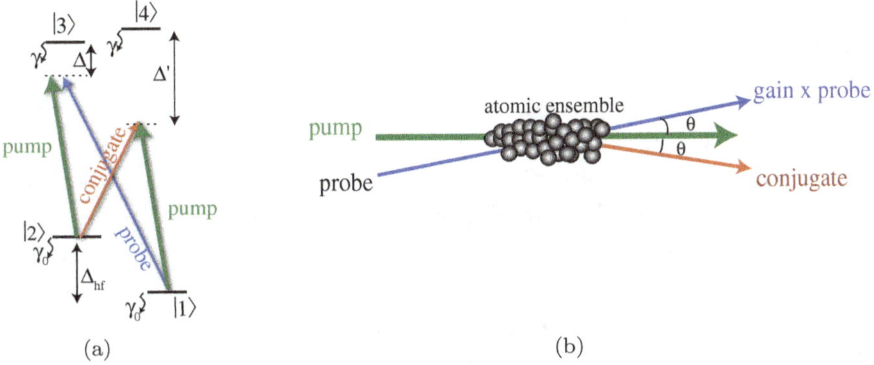

(a) A example of four-wave mixing in a four-level atom. (b) Schematic setup showing the propagation of the pump, probe and conjugate to/from an atomic ensemble. The probe and conjugate beams symmetrically propagate away from the pump beam. This is to satisfy the phase matching condition between the beams as they have slightly different frequencies.

beams with a four-level atom giving rise to generation of a fourth beam, a process known as four-wave mixing.

Consider the level structure depicted in part (a) of the following figure, where a pump field with Rabi frequency Ω and a probe field of amplitude $\mathcal{E}_p$ interact with the ground state $|1\rangle$ and metastable state $|2\rangle$ of an atom with (hyperfine) splitting frequency Δ_{hf}. Both fields are on Raman resonance and are detuned from the excited states $|3\rangle$ by amount Δ. The pump also interacts with the other ground state and generates a conjugate field, $\mathcal{E}_c$. The second two-photon or Raman transition is detuned from the excited state $|4\rangle$ by an amount Δ'. The interaction Hamiltonian of the system in a rotating frame and using RWA is given by

$$\hat{H}_{int}/\hbar = -g\mathcal{E}_p\hat{\sigma}_{31} - \Omega\hat{\sigma}_{32} - g\mathcal{E}_c\hat{\sigma}_{42} - \Omega\hat{\sigma}_{41} + h.c.$$

Here we assume equal coupling for probe/conjugate and pump transitions. h.c. refers to hermit conjugate terms.

Similar to before, we can write equations of motion for the non-negligible expectation values as

$$\frac{d\sigma_{21}}{dt} = -\gamma_0\sigma_{21} + i\sigma_{31}\Omega + +i\sigma_{42}\Omega$$

$$\frac{d\sigma_{31}}{dt} = -(\gamma + i\Delta)\sigma_{31} + ig\mathcal{E}_p + i\Omega\sigma_{21}$$

$$\frac{d\sigma_{42}}{dt} = -(\gamma + i\Delta')\sigma_{41} + i\Omega\sigma_{12} + ig\mathcal{E}_c$$

where γ is the decay rate of the excited state for both states 3 and 4 and γ_0 is the ground state decoherence or damping rate. We can see that coherence terms σ_{42} and σ_{31} are coupled via the ground state coherence σ_{12}. As the result the corresponding fields emitted at $|4\rangle \to |2\rangle$ and $|3\rangle \to |1\rangle$ transition frequencies are correlated. For this reason, the FWM process is used to generate correlated photons or squeezed or entangled states of light at the probe and conjugate frequencies.

2.9.4. *Interaction of Atoms with Light Inside an Optical Resonator*

We now consider light-atom interactions inside an optical resonator (see the figure above). We take the intra-cavity field to be single-mode at frequency ω_c represented by ladder operators $\hat{c}$ and $\hat{c}^\dagger$ and a two-level atom with energy states $|g\rangle$ and $|e\rangle$ and transition frequency ω_a interacting with the optical field with the characteristic rate g_0 (vacuum Rabi coupling rate). The atom's total decay rate is γ and the photon decay rate out of the optical cavity is given by κ. The Hamiltonian of such a system using rotating wave approximation

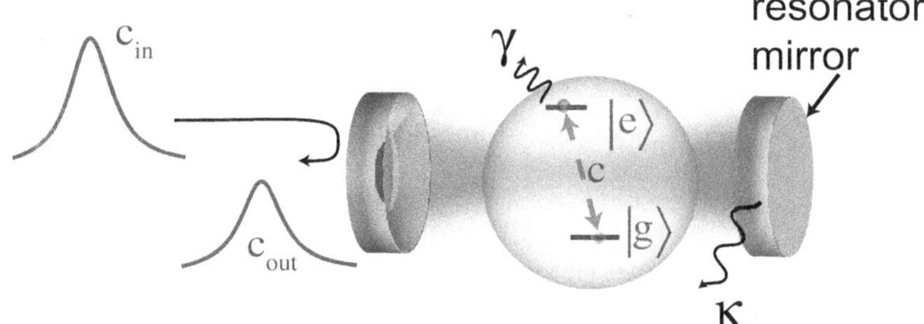

Schematics of light interacting with a single atom inside an optical resonator or cavity.

(RWA) is given by

$$\hat{H}/\hbar = \omega_c \hat{c}^\dagger \hat{c} + \omega_a \hat{\sigma}_{ee} + g_0 \hat{\sigma}_{eg} \hat{c} + g_0 \hat{\sigma}_{ge} \hat{c}^\dagger.$$

In the Heisenberg picture, the equations of motion for the optical and atomic fields become

$$\dot{\hat{c}} = -\left(i\omega_c + \frac{\kappa}{2}\right)\hat{c} - ig_0 \hat{\sigma}_{ge}$$

$$\dot{\hat{\sigma}}_{ge} = -\left(i\omega_a + \frac{\gamma}{2}\right)\hat{\sigma}_{ge} + ig_0\left(\hat{\sigma}_{ee} - \hat{\sigma}_{gg}\right)\hat{c}.$$

In the limit of low excited state population, i.e. $\langle\hat{\sigma}_{ee}\rangle < \langle\hat{\sigma}_{gg}\rangle$, we have $\langle\hat{\sigma}_{gg}\rangle - \langle\hat{\sigma}_{ee}\rangle = 1$ (because $\langle\hat{\sigma}_{gg}\rangle + \langle\hat{\sigma}_{ee}\rangle = 1$). Rewriting the equations in the rotating frame oscillating with frequency ω_c and replacing operators in the above equations with the expectation values, $\langle\hat{\sigma}_{ge}\rangle = \sigma_{ge}$ and $\langle\hat{c}\rangle = c$ we have

$$\dot{c} = -\frac{\kappa}{2}c - ig_0\sigma_{ge}$$

$$\dot{\sigma}_{ge} = -\left(i\Delta + \frac{\gamma}{2}\right)\sigma_{ge} - ig_0 c.$$

We can solve the second equation in steady state for σ_{ge} and replace it in the first equation to find

$$\dot{c} = \left(-\frac{\kappa}{2} + i\frac{g_0^2}{\Delta - i\gamma/2}\right)c.$$

The imaginary component of the second term suggests a frequency shift of the cavity resonance due to light interaction with the atom. At $\Delta = 0$, the cavity linewidth $\kappa/2$ is modified by a factor of $(1 + C)$ where $C = 4g_0^2/(\kappa\gamma)$ is the light-atom cooperativity.

We can also solve for σ_{eg} by plugging the field expectation value into the $\dot{\sigma}_{eg}$ equation and find the following dynamics at $\Delta = 0$:

$$\dot{\sigma}_{ge} = -\frac{\gamma}{2}(1 + 4g_0^2/\kappa\gamma)\sigma_{ge}.$$

The preceding equation suggests an atom decay rate modified by a factor of $(1 + C)$. This rate modification is only observable when cooperatively approaches unity or above. In this regime, emitted field from the atom can interfere with itself and modify the emission rate for the cavity more, i.e. mode $\hat{a}$.

When considering the incident and reflection light outside the cavity, we can use the following input-output relationship

$$c_{out} + c_{in} = \sqrt{\kappa}c$$

where we can define the reflection coefficient as

$$r = \frac{c_{out}}{c_{in}} = \sqrt{\kappa}\frac{c}{c_{in}} - 1.$$

Using the equation for field dynamics in the previous page at $\Delta = 0$, we can find

$$r = \frac{1}{1 + C} - 1.$$

Above, we assume the cavity is single-sided (the second or right mirror is highly reflective) and output light from the cavity travels in the opposite direction as c_{in}.

We can then derive the cavity reflection probability using the equation for field dynamics in the previous page at $\Delta = 0$, and find

$$P_r = |r|^2 = \frac{C^2}{(1 + C)^2}.$$

It can be seen that when $C \gg 1$, all of the light can be reflected with no light lost to free-space scattering or spontaneous emission.

When N identical atoms cooperatively interact with light, one can show that effective cooperativity scales with N, i.e. $C \to NC$. When atoms are not located at the maximum intra-cavity intensity, the reduced cooperativity needs to be calculated accordingly by finding the field amplitude or intensity at the location of the atom.

Chapter 3

Quantum Communication

Secure communication may play an important role in national security, financial markets and more. The need for a quantum optical communication network has been highlighted by recent demonstrations and blueprints by, for example, the Chinese Satellite Communication effort, the U.S. Department of Energy Blueprint for a Quantum Internet, Europe's Quantum Internet Alliance, and Canada's Quantum Encryption and Science Satellite, to name a few. Efficient and deterministic distribution of quantum entanglement is the key to developing future quantum networks. A network of this kind has applications in networked sensing for global parameter estimation, secure communication, and distributed quantum computing. Developing the optical interface is the only known approach to long-distance communication of quantum information.

The DARPA Quantum Network project (2000) was a landmark project aiming to push the then state-of-the-art quantum optical technology to establish the first quantum key distribution (QKD) network. Figure 3.1 illustrates the timeline and major milestones in classical and quantum technologies and indicates how research on quantum networks here is timely.

Although quantum networks have great potentials, it is difficult to predict how well they can be implemented in practice. Can a quantum network distribute entanglement over long distances and at high rates? What are the immediate applications of such networks apart from secure key distribution? What does it take to achieve distributed quantum computing?

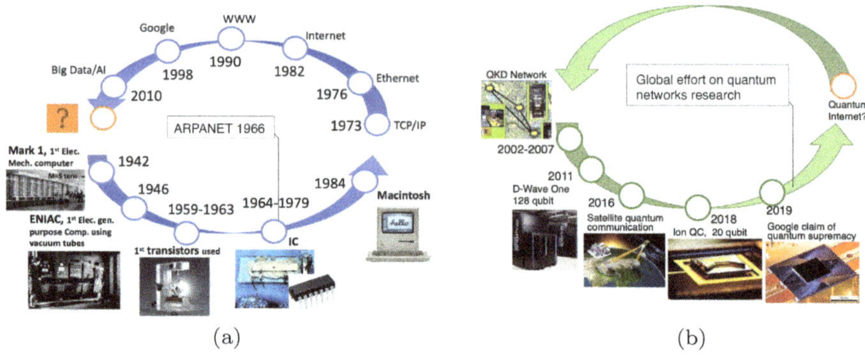

(a)   (b)

Figure 3.1. (a) Illustration of the digital technology revolution and its timeline. (b) Illustration of the next "quantum" revolution marking already-demonstrated milestones.

The main challenges preventing the community from going beyond QKD include the following:

- the probabilistic nature of photon–photon and photon–atom entangling interactions,
- the lack of scalability and multiplexing quantum communication devices,
- the incompatibility of devices in a network and the need for low-loss heterogeneous integration.

In this chapter, we introduce some quantum communication protocols and discuss how they can be implemented. We discuss the desired device metrics, device integration, and provide examples of state-of-the-art physical layers of a quantum communication system.

3.1. Quantum Key Distribution

None of the classical encryption methods proposed to date are considered entirely secure. RSA encryption is secure so long as there is no efficient way to factorize large numbers. In contrast, the laws of quantum physics enable fundamentally secure encryption.

In this section, we introduce one of the most exercised concepts in quantum communication known as quantum key distribution (QKD). QKD enables the creation and distribution of random keys between parties that are then used for encoding and decoding at those end stations. Importantly, the key distribution protocol enables the detection of eavesdroppers in the line. This is possible thanks to the no-cloning theorem, making QKD fundamentally secure from eavesdropping attacks.

In this section, we describe the well-known QKD protocol known as the BB84 protocol. We note that BB84 is the first protocol and is relatively simple to understand, but it is not certainly the only protocol used in QKD. A body of literature and experimental work is available, which relies on modified protocols to improve the security and performance of a communication system. A survey of some of the recent QKD protocols is provided in Nurhadi and Syambas, 2018.

3.1.1. *Binary Representation*

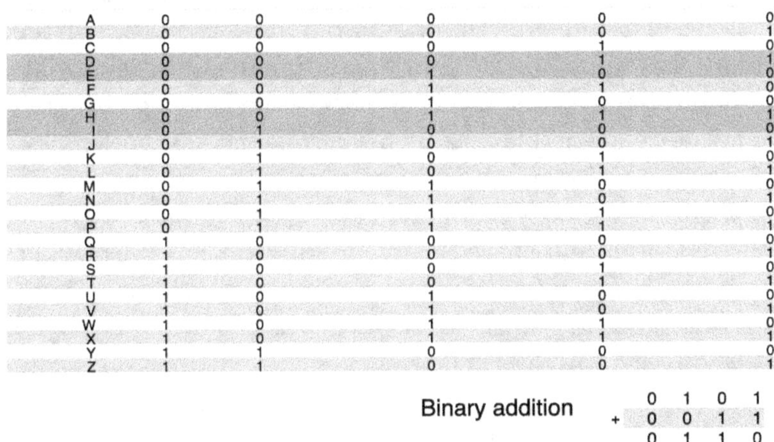

A	0	0	0	0	0
B	0	0	0	0	1
C	0	0	0	1	0
D	0	0	0	1	1
E	0	0	1	0	0
F	0	0	1	0	1
G	0	0	1	1	0
H	0	0	1	1	1
I	0	1	0	0	0
J	0	1	0	0	1
K	0	1	0	1	0
L	0	1	0	1	1
M	0	1	1	0	0
N	0	1	1	0	1
O	0	1	1	1	0
P	0	1	1	1	1
Q	1	0	0	0	0
R	1	0	0	0	1
S	1	0	0	1	0
T	1	0	0	1	1
U	1	0	1	0	0
V	1	0	1	0	1
W	1	0	1	1	0
X	1	0	1	1	1
Y	1	1	0	0	0
Z	1	1	0	0	1

$$\text{Binary addition} \quad \begin{array}{cccc} 0 & 1 & 0 & 1 \\ +\ 0 & 0 & 1 & 1 \\ \hline 0 & 1 & 1 & 0 \end{array}$$

A bit is the fundamental unit of information. Binary numbers, 0 and 1, represent the two states of a bit. A message such as a letter or a word can be written in terms of a set of binary numbers. The above table shows the binary representation of the English alphabet where each letter is represented by a set of five binary digits. As a reminder, the sum rules for binary digits are also shown above.

Using a binary operation, a message or a letter can be encoded or decoded when added to a key that is also given in the binary format.

3.1.2. *Cryptography*

Word	T	E	S	T
Binary Massage	10011	00100	10010	10011
Key	11010	10001	10100	11101
Encrypted Word	01001	10101	00110	01110
Key	11010	10001	10100	11101
Binary Massage	10011	00100	10010	10011
Word	T	E	S	T

A simple encoding process of a message "Test" using a random key is shown above. After writing the message in binary format, a key is needed which is as long as our message. Adding the key to the original message returns our encrypted message. If only the sender and receiver of the message possess the key, they can securely communicate via a public channel. Sharing the key between sender and receiver is therefore the critical part of secure communication. The key can be a set of random binary digits generated by a random number generator and shared between trusted parties. However, the key may be copied by a third party during the sharing if the key distribution is done using classical methods. The random number generators based on classical principles are also not quite random and their outcome may be predictable. The more random the key, the more secure the encryption is.

Quantum mechanics offers a new method of key distribution whereby exploiting vacuum fluctuations, true randomness can be introduced to the process from the point of view of an eavesdropper. The perceived randomness and the fact that copying quantum information is prohibited makes such communication fundamentally secure from eavesdropping attacks. In the following, we consider the realization of a basic quantum key distribution protocol using optical sources, single-photon detectors and other components.

3.1.3. *Quantum Key Generation Using Photons*

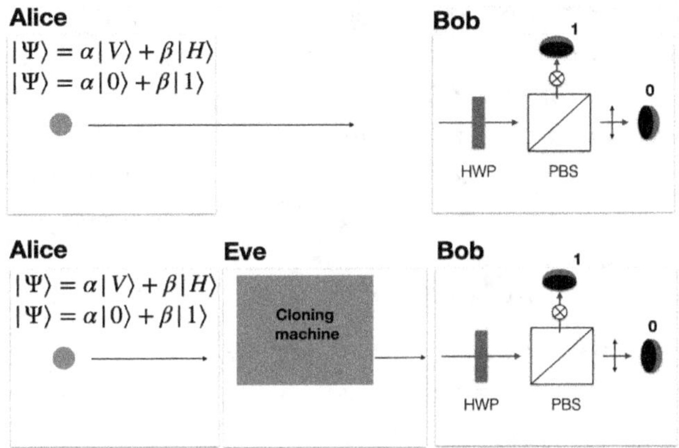

Consider the top image first where the sender Alice prepares a photon in a superposition of two polarization states V and H. We can associate V to the binary number 0 and H to the binary number 1. Alice transmits her photons to Bob, who can detect the polarization using a half-wave plate (HWP), polarizing beam splitter and a pair of single-photon detectors. If Bob sets the angle of HWP to 0 degrees, i.e. does not change the photon polarization as prepared by Alice, either detector can click. If the top detector clicks, the photon state collapses to a vertical polarization, and if the bottom detector clicks, a horizontal photon polarization is registered. In this scenario, Bob will encounter an error in the message as his measurement contains partial information sent by Alice. Instead of superposition states, Alice can send photons with V or H. In this case, Bob will be able to receive the same information sent by Alice if the HWP is set to 0 angles and the communication channel between the two does not introduce any loss or polarization rotation.

Now, consider an eavesdropper, Eve, who intercepts the channel to detect the photon sent by Alice. She tries to reproduce Alice's photon based on her detection results and subsequently sends the cloned signal to Bob. Eve does not know how Alice has prepared her state, so with a certain probability she will measure the correct state. She can then produce a photon with a polarization matching that of the measured photon and send that to Bob. In this scenario, there is a relatively high probability (50%) that Alice and Bob's photons' polarizations do not match. Alice and Bob can use a subset of the key bits to check for errors in the channel and detect the presence of Eve. In the following, we will formalize the communication and error detection protocol, as first introduced by Bennett and Brassard in 1984 (BB84).

3.1.4. *Polarization Basis*

In the QKD BB84 protocol, two polarization bases are used. We first describe how the two polarization bases are used to generate and detect photon states. The generic photon polarization state prepared by Alice can be written as $\alpha|V\rangle + \beta|H\rangle$. If either α or β is zero, the photon is horizontally or vertically polarized. This serves as one basis for photon polarization where one polarization can represent 0 and the other orthogonal polarization 1. When the photon polarization is prepared as an equal superposition state, i.e. $\alpha = \pm\beta$, the polarization is diagonal. This new polarization basis can also be used to encode 0 and 1 (see figure below).

In figure above, $\oplus$ basis refers to vertical or horizontal polarization when polarizer is set at 0 degree, and $\otimes$ basis refers to diagonal or anti-diagonal polarization when polarizer is set at 45 degrees.

For example, the state $1/\sqrt{2}|V\rangle + 1/\sqrt{2}|H\rangle$ can define bit 0 and its corresponding orthogonal state $1/\sqrt{2}|V\rangle - 1/\sqrt{2}|H\rangle$ can define bit 1. To detect the photon polarization on the diagonal basis, the HWP in Bob's station can rotate the polarization by 45 degrees to change the photon's state (via unitary transformation) to $|V\rangle$ and

$|H\rangle$ polarization. A PBS can then separate the path for $|V\rangle$ and $|H\rangle$ polarized photons, subsequently detected by a single-photon detector in each path. The generation and detection in both "plus" ($|V\rangle$ and $|H\rangle$) and "cross" ($|V\rangle \pm |H\rangle$) bases can be used to send bits 0 and 1 using two different bases. The choice of basis at Alice and Bob's locations can be random to increase the security of the key distribution.

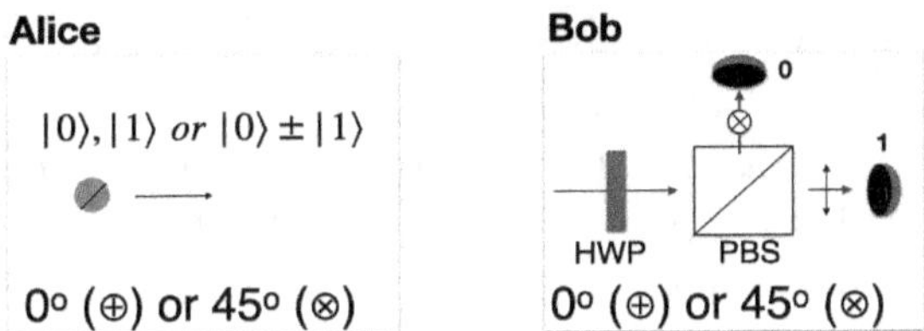

3.1.5. *BB84 Realization Using Polarization Qubits*

We now follow the recipe as proposed by BB84 protocol to see how QKD can generate a secure key between Alice and Bob. Alice decides to send a string of 0 and 1 bits by randomly choosing her polarization basis. For example, she can decide to send 0 in either "plus" or "cross" basis by rotating her photon polarization to the vertical or diagonal direction, respectively. Similarly, she can send bit 1 by rotating her photon polarization to obtain horizontal or anti-diagonal polarization. She may toss a coin or use a quantum random number generator to randomize her decision of bases. Bob on the other side of the channel receives photons and randomly decides in what basis to detect the photons. If the preparation basis of Alice happens to be the same as the detection basis of Bob, Alice's and Bob's bits will match. This is assuming there is no channel loss or noise. After a string of bits is transmitted, Alice and Bob can communicate their bases via a public channel and using classical signals. They then keep the bit results for time bins where same measurement bases were used and discard the rest. In this way, Alice and Bob can share a random key which can be used to encode and decode messages sent via a classical channel.

Alice's bit	0	0	0	1	1	1	0	0	1	0	0	1
Alice's basis	⊕	⊗	⊕	⊗	⊗	⊕	⊕	⊗	⊕	⊗	⊗	⊕
Angle(°)	90	135	90	45	45	0	90	135	0	135	135	0
Bob's bit	⊕	⊗	⊕	⊕	⊗	⊕	⊗	⊕	⊕	⊗	⊕	⊕
Bob's result	0	0	0	0	1	1	0	1	1	0	1	1
Same basis?	Y	Y	Y	N	Y	Y	N	N	Y	Y	N	Y
Key	0	0	0	↑4	1	1	↑7	↑8	1	0	↑11	1

Discarded bit# as communicated by classical channel

To determine whether Eve intercepts the signal during the bit transmission stage, Alice and Bob can reserve a fraction of their key bits and share it via the public channel to estimate the error probability. If the error is less than 25% they can assume the remaining bits are secure. This is because the presence of Eve can introduce a minimum error of 25%. Eve always has a 50% chance of guessing the preparation basis used by Alice. After detection, she needs to prepare another photon that is a copy of Alice's. However, a single detection event by Eve does not determine the state of Alice's photon. So Eve must take another 50% chance when reproducing the photon and sending it Bob. Therefore, the total error she introduces to the channel must be at least 25%.

$$P_{\text{error}} = P_{\text{Eve has wrong measurement basis}}$$

$$\times P_{\text{Bob chooses wrong preparation basis}} = 50\% \times 50\%.$$

Quantum keys can be distributed through optical fibers, free space, or satellites (Liao *et al.*, 2017). However, the communication rate is limited by how fast the photons can be created and detected, and the noise and loss of the channel (Tomaello *et al.*, 2011). Moreover, entangled photons can be used to increase the security of a channel and reduce the error probability (Wengerowsky *et al.*, 2018). It was recently demonstrated that quantum memories can help improve the QKD rate (Bhaskar *et al.*, 2020). The BB84 protocol above is described in the context of discrete variable states where single-photon detection is used to determine the presence or absence of a photon in a particle path or polarization. Continuous variable states can also be used for QKD (Qian *et al.*, 2011) where the amplitude or phase of the EM field and their quantum fluctuations can be used for the encoding of information. In this case, instead of single-photon detection, homodyne detection is used. For a summary of more recent QKD protocols that avoid some of the security loopholes of early protocols, see Pirandola *et al.* (2020).

We note that at present time, the U.S. National Security Agency (NSA) has discouraged the use of QKD for the following reasons:

- Quantum key distribution is only a partial solution.
- Quantum key distribution requires special-purpose equipment.
- Quantum key distribution increases infrastructure costs and insider threat risks.
- Securing and validating quantum key distribution is a significant challenge.
- Quantum key distribution increases the risk of denial of service.

3.2. Quantum Teleportation

One of the most exotic applications of entanglement is in the teleportation of quantum states. Teleportation relies on entanglement to remotely prepare an arbitrary quantum state. The application of teleportation is in both communication and computation. Once entanglement between two nodes is established, measurement on the part of the entangled states can teleport the quantum state to another node. The idea of distributed quantum computing relies on gate teleportation to connect multiple quantum processors and enhance computational power.

A critical building block of a distributed quantum application is a distributed quantum gate (e.g., a distributed controlled-NOT gate). Distributed gates may be implemented through the physical transport of qubits between nodes, state teleportation, gate teleportation, or protocols, such as the cat-entangler or cat-disentangler. When quantum processors are non-local, the communication between them can happen by optical qubits and optical entanglement.

In this chapter, we discuss the basic idea of quantum teleportation using optical states and particular optical measurement techniques to realize quantum optical teleportation. We note that any degree of freedom of a qubit that are entangled can be used to implement teleportation. For simplicity, we mainly focus on teleportation using polarization states of light.

3.2.1. *Teleportation Using Polarization Qubits*

Before we describe teleportation using different polarization states of light, we need to review the function of polarizing beam splitters (PBS) and define the concept of a Bell-state measurement (BSM). The left figure above shows multiple ways a PBS can output two linearly polarized fields. A PBS transmits p-polarized light while reflecting s-polarized light. In the picture above, when polarization is into the plane (in this case V-polarization) and parallel to the interface between two halves of the PBS, light is reflected (s-polarization). The orthogonal polarization, H, (or p-polarization) is transmitted through the PBS.

Now, consider the interference scheme shown in the top-right figure. This interference followed by the detection of two photons prepared in the polarization basis is an example of a BSM that can be used to teleport a polarization qubit. The yellow beam splitter (BS) is a 50/50% polarization-insensitive splitter. The BS causes the two photons to interfere if they have the same polarization. Moreover, the two photons only interfere if they spatially, temporally, and spectrally overlap. Since in total there exist two input photons, only two out of the four single-photon detectors register a click.

If one photon is V and the other is H-polarized, with a 50% probability, each will be reflected or transmitted at the BS. After the BS, the total state of the two incident photons can be written in an entangled state, e.g. $|VH\rangle + |HV\rangle$. Such a state will lead to joint clicks on detectors $D1$ and $D2$ or $D3$ and $D4$. If the entangled states are instead $|VH\rangle - |HV\rangle$, detectors $D1$ and $D4$ or $D2$ and $D3$ will register a joint click. Note that if both incident photons have the same polarization before the BS, the interference will cause both photons to go to the upper or lower path and joint detection will not be registered.

Before the BS, a polarization rotation can be applied to photons to enhance or suppress the probability of joint clicks for certain states. Working our way backward, a joint click on certain detectors will project the state of the two photons into certain entangled states.

In general, the BSM can be set up to detect two kinds of entangled states out of four possible entangled states between two qubits (Bell states): $|VH\rangle \pm |HV\rangle$ and $|VV\rangle \pm |HH\rangle$. Note that these states are not normalized and we dropped the factor $1/\sqrt{2}$.

Using the example above, we can now describe the teleportation in polarization basis. We will see how entangled qubits can be used to mediate teleportation of a third qubit via Bell state measurements.

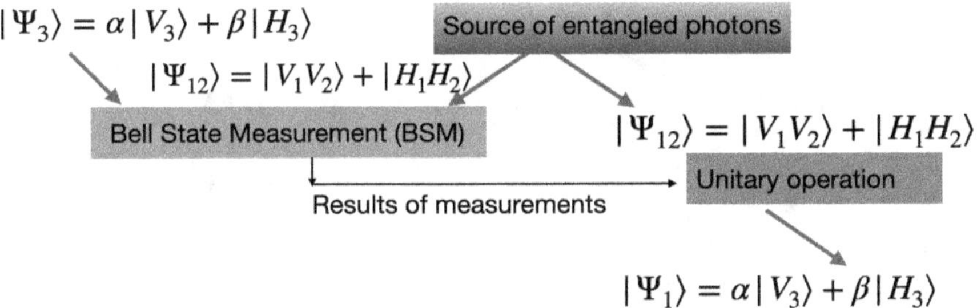

Quantum optical teleportation of polarization qubits represented by $|\psi_3\rangle = \alpha|V_3\rangle + \beta|H_3\rangle$ requires a pair of entangled photons in the same basis. An entangled state of qubits 1 and 2 represented by the quantum state $|\psi_{12}\rangle = \alpha|V_1 V_2\rangle + \beta|H_1 H_2\rangle$ describes the states of two photons spatially separated. One of the photons, e.g. photon 2, interfered with photon 3 to perform BSM. The other half of the entangled state, e.g. photon 1, is transmitted to the node where teleportation is needed. The BSM results in the projective measurement of photon 2 and photon 3. As a result, $|\psi_{12}\rangle$ describing the state of photon 1 is reduced to a single-photon state. Depending on the result of the BSM, a particular rotation is performed on photon 1 such that the probabilities α and β present in $|\psi_3\rangle$ is mapped to the state of photon 1. This completes the teleportation of information from photon 3 to photon 1 without directly sending photon 3 through the channel.

Although the effect of entanglement and projective measurement is non-local, for information to be teleported from photon 3 to photon 1, communication of BSM's result via a classical channel is necessary. Since the speed of a classical communication channel is limited by the speed of light, the teleportation of information cannot happen faster than the speed of light. When photons 2 and 3 are spectrally, temporally, and spatially mode-matched, they can interfere and subsequent detections of the photons projects photon 1 into the desired state. The result of BSM of photons 2 and 3 can produce four possible outcomes known as maximally-entangled Bell states.

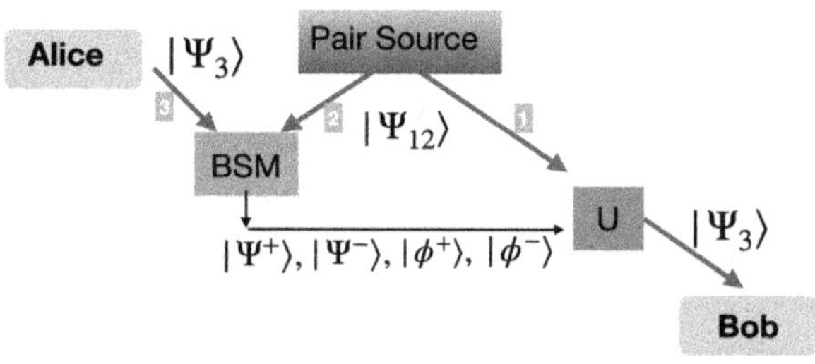

$$|\Psi^+\rangle = \frac{1}{\sqrt{2}}\left(|V_2 V_3\rangle + |H_2 H_3\rangle\right)$$

$$|\Psi^-\rangle = \frac{1}{\sqrt{2}}\left(|V_2 V_3\rangle - |H_2 H_3\rangle\right)$$

$$|\Phi^-\rangle = \frac{1}{\sqrt{2}}\left(|V_2 H_3\rangle - |H_2 V_3\rangle\right)$$

$$|\Phi^+\rangle = \frac{1}{\sqrt{2}}\left(|V_2 H_3\rangle + |H_2 V_3\rangle\right)$$

The joint states of particles 2 and 3 can take the form of one of the Bell states, as shown above. However, the BSM typically can identify three states out of four and it cannot distinguish between two of the states. In practice, using linear optical elements (e.g. beam splitters and polarizers), it is not possible to unambiguously discriminate the four Bell states with a probability higher than 50%.

After interference, using the setup shown in the figure above, the state of the three photons can be written as

$$|\Psi_{123}\rangle = \frac{1}{\sqrt{2}}[\alpha|V_1 V_2 V_3\rangle + \beta|H_1 H_2 H_3\rangle + \alpha|H_1 H_2 V_3\rangle + \beta|V_1 V_2 H_3\rangle].$$

If the BSM result identifies the following Bell state,

$$|\Psi^-\rangle = \frac{1}{\sqrt{2}}(|V_2 V_3\rangle - |H_2 H_3\rangle),$$

the state of particle 1 is then projected to

$$|\Psi_1\rangle = \langle\Psi^-|\Psi_{123}\rangle = \frac{\alpha}{2}\langle V_2 V_3|V_1 V_2 V_3\rangle - \frac{\beta}{2}\langle H_2 H_3|H_1 H_2 H_3\rangle + \cdots$$

$$= \alpha|V_1\rangle - \beta|H_1\rangle \text{ (after renormalization)}.$$

A rotation operation which is a kind of unitary operation (i.e. reversible operation that does not introduce loss or noise) can then transform the state of $|\Psi_1\rangle$ to $\alpha|V_1\rangle + \beta|H_1\rangle$, which is identical to the state of photon 3. This example shows how Alice can teleport her state to Bob without directly sending the state to him using a shared entangled state and a classical channel to communicate the results of the BSM.

3.2.2. *Experimental Quantum Teleportation*

The first experimental demonstration of optical teleportation was performed in 1997. The experimental setup is shown above. A nonlinear crystal is pumped with a UV pulse to generate a pair of down-converted red photons. The pump has reflected the crystal to generate another pair of photons propagating in the opposite

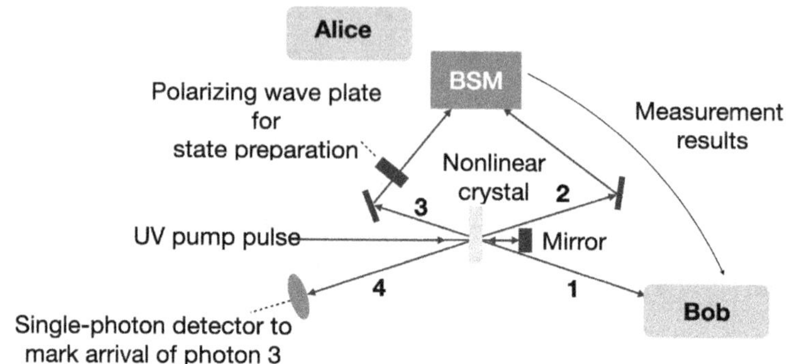

direction. The four photons generated in this way are polarization-entangled (each pair) and are mode-matched with each other. Photon 4 is detected by a single-photon detector to determine the presence of photon 3 in the upper path.

As the spontaneous down-conversion process is probabilistic, a successful detection event for photon 4 will let Alice know that her state is ready for teleportation. The remaining process is similar to what was described on the previous page. A successful BSM followed by a unitary operation at Bob's location completes the teleportation of Alice's qubit to Bob.

Note also that the detection of photon 4 indicates the presence of photon 3, it does not determine whether photons 1 and 2 exist in the other paths or not. The synchronization of photons 2 and 3 can then be a challenge limiting the success probability of teleportation. Ideally, a quantum memory can be used to synchronize photons generated by spontaneous processes. Since the demonstration of the first quantum teleportation scheme, more exotic quantum states of light have also been teleported (Lee *et al.*, 2011).

3.3. Quantum Optical Memory

Coherent mapping and transduction of quantum information carried by flying qubits (photons) to/from stationary qubits (e.g. atoms, ions, superconducting qubits) is a key step in connecting quantum devices in a network setting. Building a multi-layer quantum optical network requires quantum memories as its building blocks. Please refer to Lei *et al.* (2023) for a recent review on quantum memory developments. The efficient mapping of quantum optical information to a quantum memory can be achieved in two ways: (1) strong nonlinear interactions with single atoms mediated by a cavity or Rydberg interactions, and (2) collective interactions mediated by an ensemble of atoms. There has been a number of experiments taking the first approach in cold atoms (Li *et al.*, 2013) and solid-state (Lukin *et al.*, 2020; Bhaskar *et al.*, 2020) platforms to realize such interactions.

There are various protocols proposed and demonstrated for light storage in an ensemble of atoms (Heshami *et al.*, 2016). We review some of these protocols in this chapter.

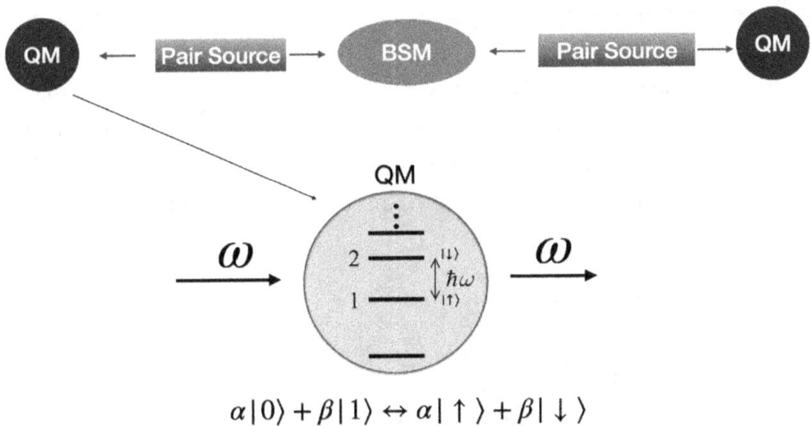

$$\alpha|0\rangle + \beta|1\rangle \leftrightarrow \alpha|\uparrow\rangle + \beta|\downarrow\rangle$$

Quantum optical states such as entangled states of light can carry and be used to distribute quantum information. However, without quantum memories, the fiber-based communication distance will be limited to <100 km due to losses in fibers. This is because transmission loss (about 0.2 dB per km) exponentially degrades the degree of quantum correlations as the communication length increases. For this reason, quantum memories were proposed as part of a "quantum repeater" system to extend the communication distance. Quantum memories storing entangled photons can themselves be entangled, thus relaying entanglement to achieve larger communication length.

The entanglement between neighboring memories can be established by Bell-state (or joint) measurement on initially unentangled photons generated by two separate entangling pair sources, as shown above.

Atoms and atomic media with suitable optical and coherence properties are attractive candidates for building quantum memories for light. To see how an atomic-based optical memory operates, consider this example; to store an optical qubit written in a superposition of two logical states 0 and 1 as $\alpha|0\rangle + \beta e^{i\phi}|1\rangle$, the amplitude and phase of the qubit should be transferred to coherence between two energy levels of an atom, e.g. $|\uparrow\rangle$ and $|\downarrow\rangle$, written as $\alpha|\uparrow\rangle + \beta e^{i\phi}|\downarrow\rangle$.

The process does not necessarily store the energy of the photon but just the information encoded in α and $\beta e^{i\phi}$. In atomic systems, the storage essentially corresponds to the coherent and reversible mapping of electromagnetic excitations. The mapping process is controlled such that atoms remember the phase of the qubit upon retrieval (phase-matching condition). The mapping from optical to/from atomic coherence should be efficient, reversible, and on demand, and the process should not add noise to retrieved optical states.

3.3.1. *Light Storage Using EIT*

We have introduced the concept of electromagnetically induced transparency (EIT) in the context of light interaction with three-level atoms. The EIT phenomenon has been suggested and used in various applications in optics and spectroscopy, such as magnetic field imaging (Mikhailov *et al.*, 2009), formation of solitons (Peng *et al.*, 2005), and coherent light storage (Fleischhauer and Lukin, 2000). Here, we describe how this phenomenon can be used to store optical information.

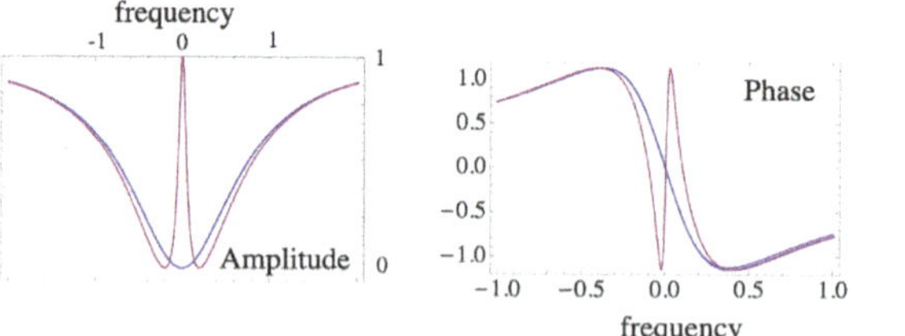

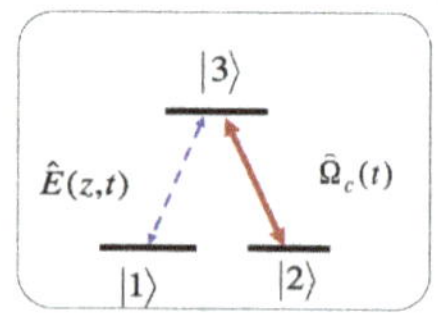

The preceding figure shows the real and imaginary parts of EIT medium susceptibility plotted as a function of frequency of the probe light relative to the atomic transition. The blue and red curves show the susceptibility resulted from interaction of probe with a two- and three-level atom, respectively. In the case of the three-level atom, a strong control light resonantly interacting with the excited state and another lower-energy state (meta-stable state) gives rise to the EIT resonance. In the EIT resonance, the imaginary part of the susceptibility is non-zero allowing the probe light to be transmitted (and not absorbed as is the case when control is not applied , i.e. when probe interacts with a two-level atom). The real part of the susceptibility is related to the refractive index of the atomic medium and its steep variation near resonance in the EIT condition, suggests high group index and low group velocity. In the previous chapter, we discussed how to derive the equation for the medium susceptibility. Thanks to quantum interference between the two absorption paths, under the EIT condition, the propagation of a weak resonant field can be controlled using a strong resonant control field. The reduced group velocity of light under this condition is associated with the creation of atomic coherence inside the medium. In this interaction regime, the system cannot be described by the individual fields associated with atoms or light. We can define a quasi-particle, known as the dark-state polariton (Fleischhauer and Lukin, 2000), to capture dynamic interaction between light and atoms.

$$\hat{\Psi}(z,t) = \cos\theta(t)\hat{E}(z,t) - \sin\theta(t)\sqrt{N}\hat{\sigma}$$

$$\cos\theta(t) = \frac{\Omega(t)}{\sqrt{\Omega^2(t) + g^2 N}}$$

$$\sin\theta(t) = \frac{g\sqrt{N}}{\sqrt{\Omega^2(t) + g^2 N}}$$

The dark state polariton, $\hat{\Psi}(z,t)$, can be written as a coherent superposition of a light field envelope, $\hat{E}(z,t) = \sum_n \sqrt{\frac{\hbar\omega_n}{2V\epsilon_0}}[\hat{a}_n^\dagger(t) + \hat{a}_n(t)]u_n(x)$ and atomic coherence, $\hat{\sigma} = \frac{1}{N}\sum_j^N |1_j\rangle\langle 2_j|$. To see how we can find θ, we can plug the solution, i.e. $\Psi(z,t)$, into the following equation of motion derived from the Maxwell and the Schrödinger equations:

$$\left(\frac{\partial}{\partial t} + c\cos^2\theta\frac{\partial}{\partial z}\right)\Psi(z,t) = 0,$$

where c is the speed of light in vacuum and Ω is the Rabi frequency of the control laser. The light's group velocity is given by $c\cos^2\theta$. It is called the dark-state polariton as due to quantum interference of absorption paths, the atoms' excited state population is small. In other words, the solution, $\Psi(z,t)$, does not include terms associated with the excited state, $|3\rangle$, population.

The ratio between the optical and atomic excitations can be written in terms of the Rabi frequency of the control field, Ω, and the effective atom-light coupling, $g\sqrt{N}$, where N is the number of atoms interacting with the light. It can be seen that the polariton can change from all-photonic to all-atomic by reducing the control field power to zero (for a portion of light already inside the atomic medium). In this case, the amplitude and phase information of the light pulse is mapped into the atomic coherences, and so, the light field is stored coherently. By increasing the control field power, the system can change to the superposition state. The polariton is reaccelerated and eventually becomes all photonic, i.e. $\cos\theta = 1$, as the light pulse leaves the medium with the vacuum speed c. For the optimum storage of a data pulse, the EIT medium is required to

$$\hat{\Psi}(z,t) = \cos\theta(t)\hat{E}(z,t) - \sin\theta(t)\sqrt{N}\,\hat{\sigma}$$

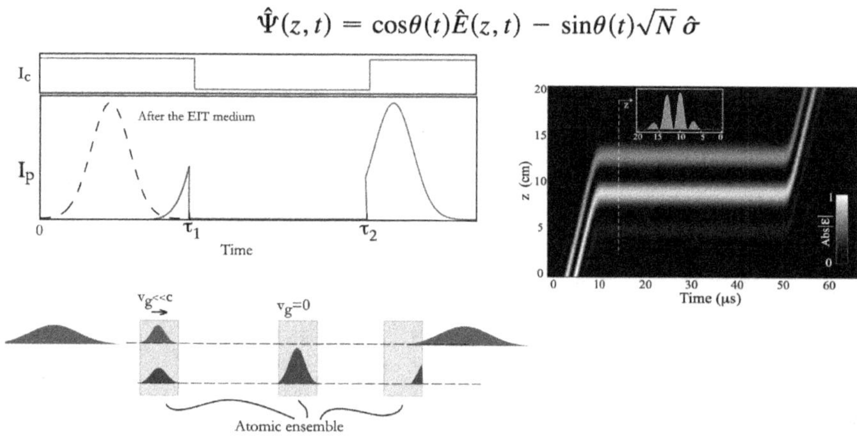

provide both a wide transparency window and a low group velocity to spatially localize the entire pulse inside the ensemble. Otherwise, a fraction of data will leave the medium without interaction after control laser is switched off. This is shown in the figure above (left) where the control field is switched off between τ_1 and τ_2 while a part of the input pulse is transmitted without storage.

The EIT polariton is depicted in the figure above (right) in the $z - t$ plane for a modulated input pulse. Before the control field is adiabatically switched to zero (t <15 μs), the system is in a superposition of the atomic and photonic fields, where the light field is propagating with a group velocity ($v_g \ll c$). The polariton intensity remains constant over time (assuming low decoehrence rate) because it accounts for both atomic and optical fields. After the control field is switched to zero, the information is mapped into the atomic coherence and the polariton becomes all-atomic. As it is shown in the inset, the temporal profile of the input pulse is obtained by taking the cross-section of the polariton along the propagation axis. By turning back on the control field, the light field is regenerated and eventually released at the end of the sample and propagates with vacuum speed, $v_g = c$. A detailed treatment of EIT can be found in the works of Fleischhauer *et al.* (2005) and Lukin (2003).

3.3.2. *Principles of Atomic Frequency Comb Memory*

A photon-echo QM based on the atomic frequency comb (AFC) method was proposed by Afzelius *et al.* (2009) and since then many light storage experiments have been performed based on this protocol. In the originally proposed atomic frequency comb technique for light storage, the inhomogeneous broadening in atomic transitions, which is present in solid-state systems, was used to generate photon echoes recovering the stored optical coherence. The inhomogeneous broadening is the result of a non-uniform frequency-shifting interaction of the optical centers (or atoms used for storage) and the environment, which is manifested as a continuous absorption spectrum, much broader than the single-atom linewidth. The AFC takes advantage of such broadening to create broadband storage.

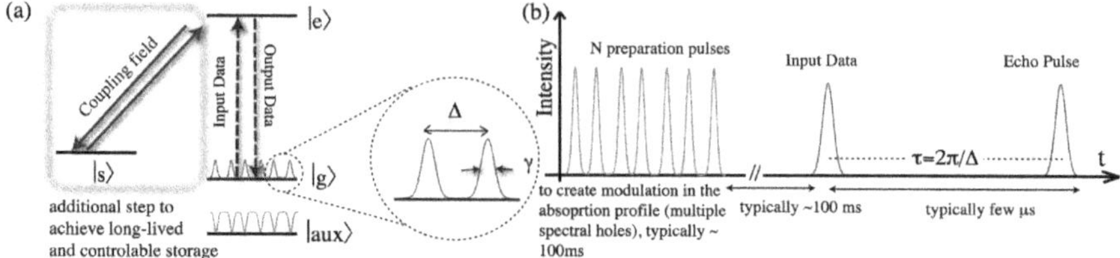

In the simplest form, a pump laser is used to "burn" absorption
holes in the broad inhomogeneous line. This is known as the spectral
hole burning technique (Trommsdorff *et al.*, 1995). As atoms with
certain frequencies matching that of pump frequency are excited,
they decay to another lower-energy state ($|g\rangle$) via spontaneous
or stimulated emission processes. The state $|g\rangle$ is ideally empty
for certain-frequency atoms and has a long coherence time. This
creates an anti-hole or enhanced absorption for light interacting
with energy state $|g\rangle$ for those atoms. Multiple holes in state $|aux\rangle$
and anti-holes in state $|g\rangle$ with width γ can be generated with
a fixed frequency spacing (Δ) by proper engineering of the pump
frequency and intensity. This creates an atomic frequency comb
that can fully absorb a broadband input pulse, see figure above.
Although the atomic spectrum after the preparation (hole burning)
consists of discrete absorption peaks separated by Δ, the energy–
time uncertainty relation gives rise to a smooth absorption of the
entire input pulse spectrum (see Afzelius *et al.* (2009) for more
details).

The process allows for a uniform mapping of a broadband light
pulse onto discrete atomic excitations.

$$|\psi\rangle = \sum_{j=1}^{N} c_j e^{i\delta_j t} e^{-ikz_j} |g_1 \cdots e_j \cdots g_N\rangle$$

where we define AFC Finesse:

$$F = \frac{\Delta}{\gamma}, \quad \text{and find AFC efficiency:} \approx \left(\frac{d}{F}\right)^2 e^{-7/F^2} e^{-d/F}$$

As the relative phase between the neighboring anti-holes matches at every $2\pi/\Delta$, a coherent photon echo is emitted to regenerate the input light. This method has been used for quantum storage of optical information with high bandwidth (Saglamyurek *et al.*, 2011; Saglamyurek *et al.*, 2015; Clausen *et al.*, 2011). The ratio of Δ to γ is called AFC finesse which together with the optical density, $d = g^2 N L/\Gamma c$ defines the efficiency of the memory. Here, L is the length of the memory and Γ is the atomic spontaneous emission rate. Furthermore, a coupling light can be used to transfer the excitation, after the data is absorbed, to another long-lived state, e.g. $|s\rangle$, to extend the coherence time that might be limited by the decay of the excited state. This additional step also enables one to control the retrieval time since no echo will be emitted so long as the excitation is contained between $|s\rangle$ and $|g\rangle$ states.

3.3.3. *Light Storage Using Gradient Echo Memory*

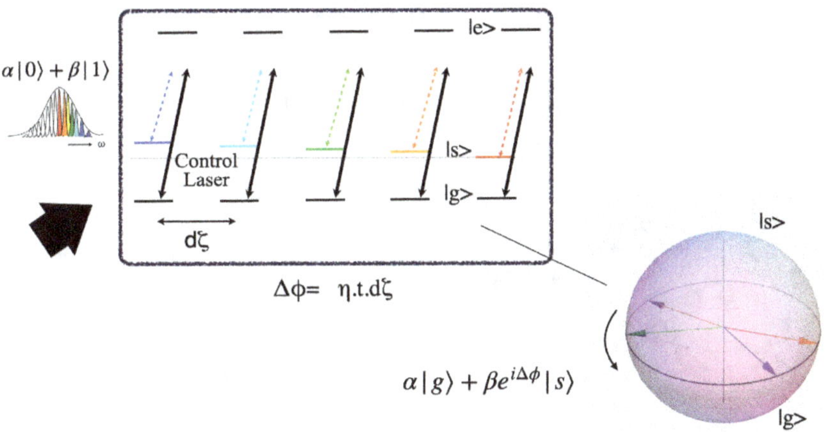

The gradient echo memory (GEM) also relies on inhomogeneous broadening, but a controllably induced one, for coherent light storage. Consider an ensemble of three-level atoms subject to an external control laser detuned from the excited state and an external magnetic or electric field gradient. The control laser enables the absorption of a portion of the input pulse whose frequency components satisfy the Raman resonance condition for certain atoms. The external field induces a Zeeman or Stark shift of the atomic energy levels that linearly changes across the sample. The gradient of frequency broadening introduced by the external field ensures that different frequency components of the photon wavepacket are stored at different locations, thus printing the phase information across spatial locations. This resembles the strings in a harp or the way the cochlea in our inner ear works.

As the frequency difference between atoms along the sample is different, the evolving phase difference prevents coherent or cooperative emission. This can be seen by looking at the Bloch sphere. Assuming the input light is strong enough, its absorption can change the atoms' state, initially prepared in $|g\rangle$, to an equal suppression of $|g\rangle$ and $|s\rangle$. The atomic spins can be represented by Bloch vectors on the equator that rotate at a different speed. So, the relative phase between the Bloch vectors increases as time elapses.

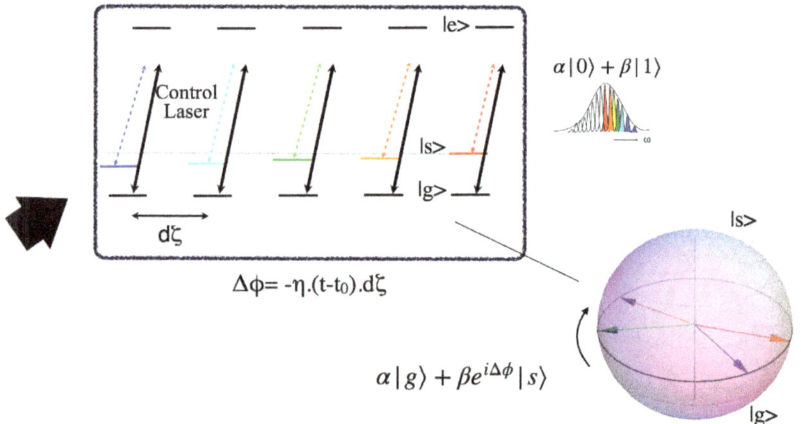

Source: PRL 101, 203601 (2008); Optics Lett. 33, 20 (2008).

Now, by reversing the slope of the gradient, the process can be effectively time-reversed. In other words, the phase change between two regions of the memory separated by $d\zeta$ is $\Delta\phi = \Delta\omega.t$, where $\Delta\omega = \eta d\zeta$ and η is the slope of the gradient. The total accumulated phase change after time t_0 can be zero if the gradient slope is reversed and it remains that way for another t_0 time. At this time, an echo emerges from the memory that is a time-reversed copy of the input light.

In this case of three-level atoms, the problem can be reduced to a two-level system, assuming the detuning of the strong control light from the excited state is large compared to the excited state linewidth. In this case, the atoms do not spend much time in the excited state to spontaneously decay, and therefore, the excited state can be adiabatically eliminated from the equations or description of the system. The interaction described above can be simplified by considering an effective field interacting with levels $|g\rangle$ and $|s\rangle$ with an effective coupling strength $g' = g\frac{\Omega}{\Delta}$, where Δ is the detuning of the Raman transition from the excited state.

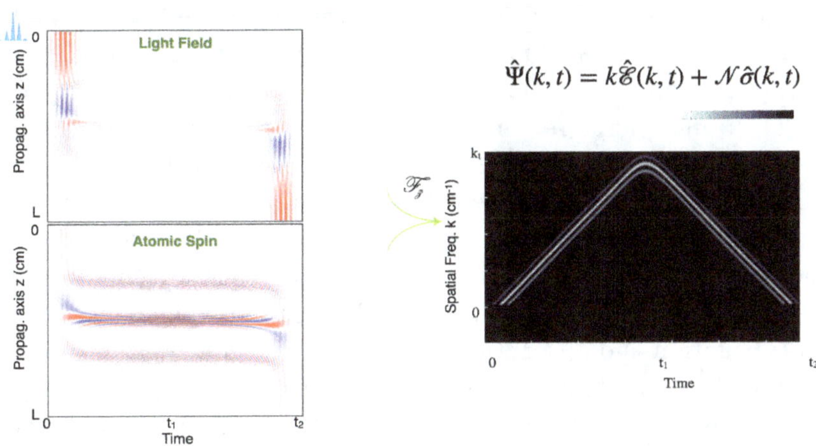

Similar to the EIT case, one can associate a quasi-particle (polariton) consisting of the light field and atomic coherence (spin wave) to the light–atom dynamic inside a GEM. In the simpler case of two-level atoms, by taking the spatial Fourier transform of the equations of motion corresponding to the atomic coherence and the optical field, we arrive at the following form for our polariton:

$$\hat{\psi}(k,t) = k\hat{\mathcal{E}}(k,t) + \mathcal{N}\hat{\alpha}(k,t),$$

which has the following equation of motion

$$\left(\frac{\partial}{\partial t} + \eta_{(t)} \frac{\partial}{\partial k} - i\frac{g\mathcal{N}}{k} \right) \hat{\psi}(k,t) = 0.$$

Here, the effective linear atomic density $\mathcal{N} = gN/c$, k is the spatial frequency component, η is the slope of gradient, and ηz is the frequency shift of the atom at location z, which appears as a detuning in the equation of motion. Taking the spatial Fourier transform, $\eta z \to \eta \partial/\partial k$, helps to define a well-defined polariton for the system in the Fourier domain.

By setting $\mathcal{N} \to \frac{\mathcal{N}\Omega}{\Delta}$, $g \to \frac{g\Omega}{\Delta}$, the equations of motion and polariton picture for a two-level atom can approximately capture the dynamics of the corresponding three-level system.

One can write another solution for the above equation of motion for a polariton as $\hat{\psi}(k,t) = k\hat{\mathcal{E}}(k,t) - \mathcal{N}\hat{\sigma}(k,t)$. However, it can be shown using the Maxwell equation that $k\hat{\mathcal{E}}(k,t)$ is equal to $\mathcal{N}\hat{\sigma}(k,t)$, and therefore, this mode will not be excited. The normal mode equation of motion indicates that $\psi(k,t)$ propagates in the k-axis with a speed defined by the slope of the gradient. As the polariton reaches higher k values, the electric field amplitude gradually decreases through time. As each frequency component of the light propagates through the medium, the dispersion will affect its propagation and absorption, but the polariton will propagate with no change in its intensity (assuming low decoherence).

Flipping the gradient (i.e. $\eta \to -\eta$) time reverses the absorption process so that when the polariton reaches $k = 0$ (the phase matching condition), a photon-echo emerges from the ensemble in the forward direction. The above simulation shows the optical, atomic, and polariton excitation for a modulated input pulse in time and space.

The group velocity of a light field propagating inside a GEM medium is given by $v_g = g\mathcal{N}/k^2$. The farther away the polariton is from $k = 0$, the less intense the electric field and also the smaller the group velocity. The spatial cross-section of the atomic polarization at any time during the storage is the Fourier spectrum of the input pulse. This explicitly demonstrates the frequency-encoding nature of GEM.

3.3.4. *Efficiency and Fidelity of a Quantum Memory or Communication Channel*

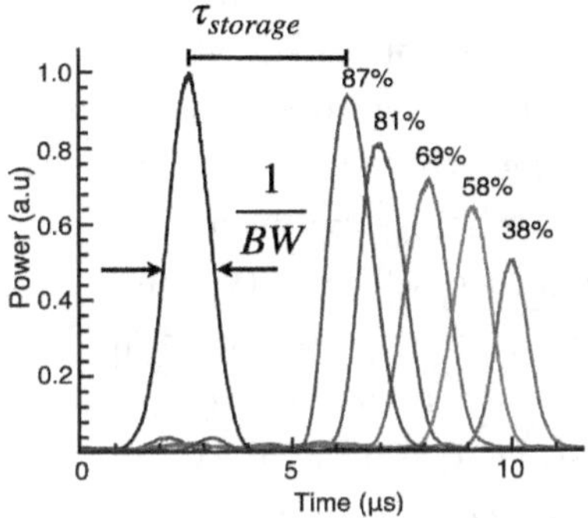

The figure above on the left shows actual experimental data for the storage of a light pulse using the GEM technique. By delaying the switching time of the memory, the echo exits the memory at a later time. This enables us to tune the storage time. The efficiency of the memory is defined as the total output optical power divided by the total input power.

$$\frac{\int P_{\text{out}}(r,t)\,dt\,dr}{\int P_{\text{in}}(r,t)\,dt\,dr}$$

Assuming the increased atomic density does not affect loss and decoherence in the medium, the efficiency can be increased by adding more atoms to the system. But as can be seen above, the efficiency drops as the storage time increases. The rate of efficiency drop is defined by the inverse coherence time of the memory. Depending on the underlying decoherence effects, the loss in efficiency may follow exponential, Gaussian, or other profiles as a function of storage time.

To quantify the noise of a quantum memory, the fidelity of input–output quantum states is measured. This can be done by reconstructing the density matrix of the input and output light pulses. Once the density matrix elements are measured, the equation above for the fidelity, $\mathcal{F}$, can be used to calculate the fidelity.

$$\mathcal{F} = |\langle \psi_{\text{in}} | \psi_{\text{out}} \rangle|^2 = \left[\text{Tr} \sqrt{ \sqrt{\rho_{\text{in}}}\, \rho_{\text{out}} \sqrt{\rho_{\text{in}}} } \right]^2$$

Fidelity is a measure of overlap between the output and input quantum states.

A quantum memory can show near-unity fidelity but with <1% efficiency or vice versa. A brick can "store" a vacuum state with 100% fidelity (vacuum in, vacuum out). If the fidelity of memory is measured only when the state is retrieved, it is the "conditional" fidelity that is being evaluated. It is therefore important to measure both conditional fidelity and efficiency to verify the memory performance in the quantum regime. The "unconditional" fidelity takes the efficiency into the account as it measures the state overlap even when storage is not successful (or the state is lost to decoherence). Although conditional fidelity can be high, low memory efficiency limits the communication rate. Moreover, the bandwidth of a memory device is another important parameter that determines the rate of communication.

3.3.5. *State Tomography*

A quantum optical memory should preserve both the wave and particle nature of the quantum information encoded into an electromagnetic field. This is quantified by, for example, measuring the density matrix of the light after performing quantum tomography.

$$|\Psi\rangle = \sum_i c_i|\phi_i\rangle \longrightarrow \hat{\rho} = |\Psi\rangle\langle\Psi| \longrightarrow \hat{\rho} = \sum_{ij} c_i c_j^*|\phi_i\rangle\langle\phi_j|$$

When the amplitude and phase of the light are used for encoding (continuous variable regime), quantum tomography can be done by performing homodyne measurements. The Homodyne measurement is an interferometric detection that returns a signal proportional to the amplitude of the quantum field as a function of the field's phase relative to a reference classical electromagnetic field. When discrete variables such as polarization or photon number states are used for encoding (discrete variable regime), a series of coincident measurements in different polarization basis is performed to evaluate the density matrix elements.

The figure above shows the reconstructed density matrix for the input and output states of an optical memory in the number or Fock basis that are eigenstates of the energy (Hamiltonian) or number operator. Here, a weak coherent state was used as the input to the memory.

$$|\psi_{\text{in}}\rangle = c_0|0\rangle + c_1 e^{i\phi_{01}}|1\rangle + \cdots$$

For example, the probabilities at $m = 1$ and $n = 1$ or $m = 2$ and $n = 2$ represent population terms or probability of having 1 or 2 photons, respectively. On the other hand, probabilities for cases where $m \neq n$ are coherences. The existence of non-zero coherence terms shows a coherent phase relationship between different photon number states at the input and output of the memory (i.e. coherent storage).

Typically, the measurements are not complete and the data do not need to exhaust all the values to have an accurate description of the density matrix or the wave function. In this case, the full quantum description of a system may be accomplished using maximum likelihood estimation (Hradil, 1997). This method enables one to perform a set of measurements on various known quantum states and then estimate the unknown in the measurement from the collected data. Reconstruction corresponds to the normalization of incompatible observations that capture part of the possible outcomes. This approach handles noisy data corresponding to a realistic incomplete observation with finite resolution. The reconstruction suggests what state best matches the measurements. This type of reconstruction can be used to estimate the quantum state of a system using, for example, homodyne measurements.

3.3.6. *Quantum Memory Platforms*

The most basic form of an optical memory is a fiber delay line. The delay in this case is mostly fixed and proportional to the fiber length. The delay can, in principle, be controlled by rapid switching of the fiber length. Compared to the atomic memories, the fiber delay lines are broadband, can operate at different frequencies near the telecom bands, and can also support multiple wavelengths. The maximum delay time in a fiber is limited by the transmission loss which is typically about 0.2 dB per km. For example, a 300-km-long fiber line can provide about a millisecond of delay while introducing 99.8% loss.

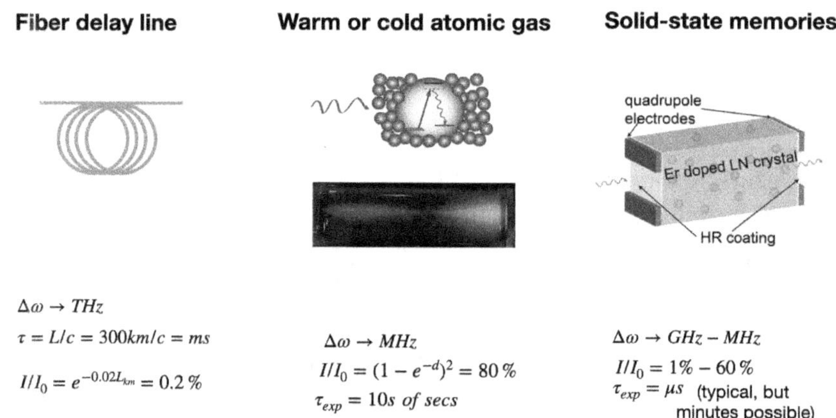

Atomic memories can be implemented in both gas and solid-state media. In this case, typically, optical coherence is transferred to atomic coherence that can, in principle, reach seconds or even minutes of storage time. The bandwidth of gas-phase atomic quantum memories demonstrated to date is on the order of MHz, although GHz-bandwidth memories can also be achieved. A storage time on the order of seconds (in laser-cooled atomic gases) along with storage efficiencies close to 80% have also been achieved. Solid-state quantum memories such as rare earth solids do exhibit large inhomogeneous broadening that can be leveraged to implement multi-GHz-bandwidth memories. Using optical cavities or high purity crystals (low inhomogeneous broadening), high storage efficiencies can also be achieved. The storage time achieved in typical demonstrations is on the order of a microsecond. Recently, coherence time of a few minutes has also been demonstrated. The storage time in atomic memories can be controlled by external electro-magnetic fields. External fields can be used to control the mapping of light to/from atomic coherence by controlling the phase-matching condition or light–atom coupling strength.

3.3.7. *Memory Compatibility and Integration*

Apart from bandwidth, efficiency, fidelity, and storage time, a quantum memory should be compatible with quantum sources, and optical infrastructure and its implementation should be scalable. For fiber-based quantum communication, both sources and memories should ideally operate near the telecom band where the fiber loss is minimum. When either wavelength is different from the telecom band, an additional process involving wavelength conversion is needed, which itself can impose additional loss, noise, and/or bandwidth limitations.

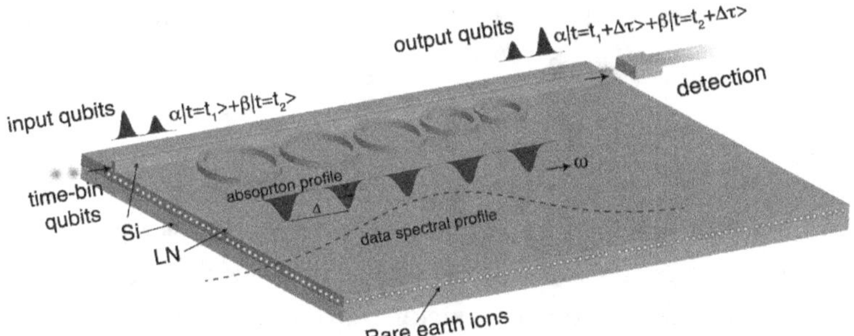

As we shall see in the following section, for practical quantum communication purposes, many quantum memories and sources are required per node. Therefore, a quantum memory with suitable properties should also be scalable to accommodate the needs of a multiplexed network. Heterogeneous or homogeneous integration of sources, memories, and other components (e.g. switches, modulators, and filters) comprise the main challenges of future quantum networks.

3.4. DLCZ Entanglement Distribution Protocol

The concept of a collective excitation in an atomic ensemble is key to many coherent processes including quantum storage (Tanji *et al.*, 2009), bi-photon generation (Thompson *et al.*, 2006), and entanglement creation (Jing *et al.*, 2019). Such collective effects when combined with projective measurements have applications in quantum communication, as proposed by Duan, Lukin, Cirac and Zoller (DLCZ) (Duan *et al.*, 2001). In this section, we describe the basic concept of a measurement-based entanglement distribution between non-local atomic ensembles, as originally proposed by Duan, Lukin, Cirac and Zoller in 2001.

3.4.1. *Duan–Lukin–Cirac–Zoller Proposal (2001)*

$$|\phi_{light}\rangle = A(|0\rangle + p|1\rangle + \frac{p^2}{\sqrt{2!}}|2\rangle + \dots)$$

$$\xrightarrow{\ \ p \ll 1\ \ } \quad |\psi\rangle = |0\rangle|g\rangle + p|1\rangle|s\rangle$$

Consider an ensemble of N_a three-level atoms (Λ system), as shown in the figure above. The state $|g_i\rangle$ represents the ground state, and $|s_i\rangle$ is the meta-stable state (another lower-energy state with a forbidden optical dipole transition and relatively long coherence time) of atom i. The state $|e\rangle$ is the excited state via which the atom is transferred between $|g\rangle$ and $|s\rangle$. The atomic ensemble consists of many identical atoms that can experimentally be realized by, for example, two hyperfine levels of the alkali atoms' ground state interacting with one of the excited states via two-photon transitions. The atomic state, $|g\rangle$, represents the collective state of all atoms when they are in the ground state. The state $|s\rangle$ represents a different collective state (a single collective excitation) when only one of the atoms is excited to the $|s_i\rangle$ state (for atom i). The summation over i denotes the sum of all permutations.

$$|g\rangle = |g_1 g_2 \cdots g_{N_a}\rangle \longrightarrow \quad |s\rangle = \frac{1}{\sqrt{N_a}} \sum_{i=1}^{N_a} |s_i\rangle \quad \text{spin wave}$$

$$= \frac{1}{\sqrt{N_a}} \{|s_1 g_2 \cdots\rangle + \cdots + |g_1 \cdots s_{N_a}\rangle\}$$

Such single excitation can be created within a large atomic ensemble by, for example, storing a single photon in an ensemble of atoms. Alternatively, a single atomic excitation can be created using an off-resonance Raman process to scatter a single photon from atoms initially in the $|g\rangle$ state. A "write" laser off-resonantly interacting with the $|g\rangle \to |e\rangle$ transition can lead to the generation of a Raman-scattered photon (Stokes photon). If only a single photon is detected within a specific spatial mode, only one atom within that spatial mode could have emitted it. Since we have no way of telling which atom, we write the atomic state as a coherent superposition of all different possibilities. The state $|s\rangle$ is representative of a single collective excitation only if the relative phase between atoms remains constant during a given timescale (coherence time). The frequency noise or motion of atoms can create phase noise implying that the state can no longer be described by the pure state $|s\rangle$.

The total state of an atom–photon system can therefore be written as an entangled state, $|\Psi\rangle$. Detecting the photon (Stokes photon) in the number basis, $|1\rangle$, projects the state $|\Psi\rangle$ to $|s\rangle$.

A single collective excitation can store the coherence between $|g_i\rangle$ and $|s_i\rangle$ (the spin wave given by $|s\rangle$) for a limited time given by the coherence time of the medium. The subsequent application of a "read" laser light that is resonant with the atomic transition $|s\rangle \to |e\rangle$ results in the reading out of the coherence in the form of an anti-Stokes photon correlated with the Stokes photon. The two read and write pumping processes are controlled by the time, power, and detuning of the lasers from the excited state. The photon generation is governed by the spontaneous Raman scattering process, whose optical state is ideally described by a coherent state.

$$|\phi_{\text{light}}\rangle = A \left(|0\rangle + p|1\rangle + \frac{p^2}{\sqrt{2!}}|2\rangle + \cdots \right)$$

The Stokes process is therefore probabilistic and follows Poisson statistics. The scattering probability of certain photon-number states, p, are proportional to the power of the write laser beam. The retrieval of the anti-Stokes photon however can be deterministically achieved via a resonant two-photon process (e.g. EIT). In the EIT process, the interference gives rise to the transparency of the medium, reducing the Raman absorption and loss of the anti-Stokes photon. Although increasing the "write" power can increase the rate of Stokes photon generation, it also increases the probability of generating multiple photons, introducing noise to the process. Therefore, the detuning and power of the "read" laser should be chosen such that the probability of generating a single photon remains low and the probability of generating more than one photon is negligible. The efficient generation of photon pairs using this Raman process has been realized using both room temperature atoms (Eisaman and *et al.*, 2005) and cold atoms in free space or optical cavities (Thompson *et al.*, 2006).

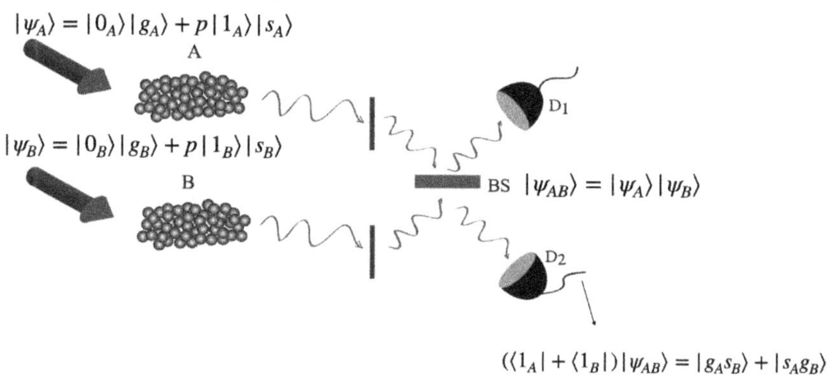

$$(\langle 1_A| + \langle 1_B|)|\psi_{AB}\rangle = |g_A s_B\rangle + |s_A g_B\rangle$$

The Stokes and anti-Stokes processes of this kind have been proposed to distribute entanglement between remote atomic ensembles (Duan *et al.*, 2001). Consider two atomic ensembles, A and B, where both are subject to two identical "write" lasers. If the two Stokes photons are probabilistically generated, each from one ensemble, the two entangled states $|\Psi_A\rangle$ and $|\Psi_B\rangle$ (states above are not normalized) can be created, where a photon A (B) is entangled with the ensemble A (B). The state of the joint system is not entangled and can be written as a product state: $|\Psi_A\rangle|\Psi_B\rangle$. Now, if either detector registers a single click after the two paths interfere, there is no way of determining if the photon was emitted from ensemble A or B. As a result, the state of the system will be projected to an entangled state between ensemble A and B, as seen above.

A more useful entangled state can be created when photons A and B are generated in the polarization basis (dual-rail qubit) instead of the number basis (single-rail qubit), i.e. $|V_{A/B}\rangle+|H_{A/B}\rangle$. In this case, the BSM between the two photons in the polarization basis (requiring coincident detection events) can project the two ensembles into an entangled state.

Now, consider four ensembles A, B, C, and D. The paths and detection scheme can be arranged such that ensembles A and B and ensembles C and D are entangled by a joint photon detection between corresponding photons (similar to what is shown above for ensembles A and B). Once A–B and C–D are entangled, subsequent retrieval of photons from ensembles B and C (using "read" lasers) and performing additional BSMs between retrieved B and C photons can "swap" entanglement, leaving us with A–D entanglement. In this way, neighboring ensembles in a channel can be entangled sequentially until the end ensembles are entangled. In certain regimes, this strategy can overcome losses in fibers where instead of exponential loss of entanglement, the losses can take the form of polynomial functions with distance.

3.4.2. *Probability and Noise*

The entanglement swapping process described on the previous page can enable the distribution of entanglement over long distances. However, there are some technical and fundamental challenges when it comes to the success probability of such a technique. The sources of failure of the process are higher-order processes or multi-photon processes, mode-mismatching loss, and spontaneous emission loss (Duan *et al.*, 2002). The noise induced by higher-order processes can be mitigated by reducing the pump power at the expense of reducing the entanglement creation rate.

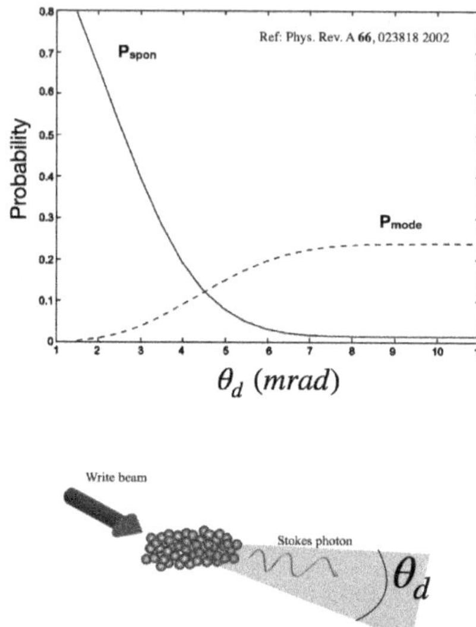

In short, there are two fundamental noise sources:

- The spontaneous emission loss represents the probability of not detecting photons while atoms are excited to a collective spin state.
- The mode-mismatching loss represents the probability of not creating an atomic excitation even though a Stokes' photon is detected.

The spontaneous emission loss (or noise) represents the probability of not detecting photons within the detection mode (with collection solid angle θ_d) even when atomic coherence is created. Similarly, the mode-mismatching loss (or noise) represents the probability of not creating an atomic excitation even when photons are detected within the detection mode.

The calculation result (Duan *et al.*, 2002) of the two noise sources as a function of θ_d is shown above. The optimum solid angle can be found where the two noise probabilities cross. The optimum point can vary with other experimental parameters such as optical density. A simplified equation for the spontaneous noise probability is written below in terms of the optical density or d at a given collection θ_d

$$P_{\text{spon}}|_{\theta_d \sim 3\,\text{mrad}} = \frac{1}{1 + d/26}.$$

3.5. Long-Distance Entanglement Distribution

Efficient and deterministic distribution of quantum entanglement is key to developing future quantum networks. A network of this kind has applications in networked sensing for global parameter estimation, secure communication, and distributed quantum computing. Developing an optical interface is the only known approach to long-distance communication of quantum information. In this section, we focus on the physical layer of such networks while paying particular attention to quantum storage, multi-mode communication, and device integration requirements.

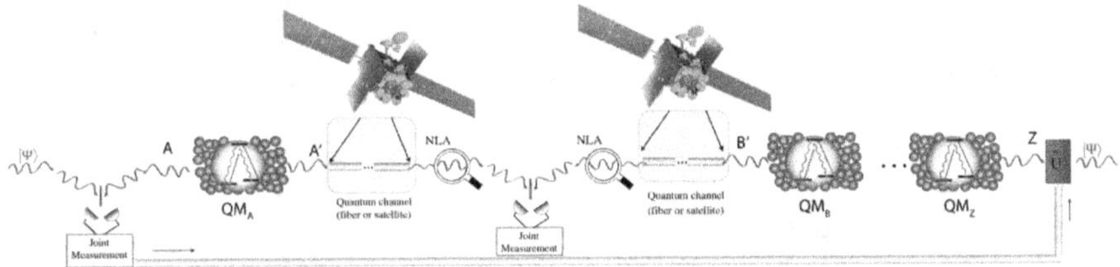

Quantum optical states such as entangled photons can carry quantum information; however, without quantum memory, the fiber-based communication distance is limited to about 100 km due to losses in the fibers, which exponentially degrades the degree of quantum correlations. For this reason, quantum memory was proposed as a component of a quantum repeater system to extend communication distances. Quantum memories storing entangled photons can themselves be entangled. As seen above, entanglement between neighboring memories can be established by Bell state (or joint) measurement on photons from nearby memories. Subsequent measurements lead to the creation of entanglement between memories in distant nodes. The information can then be teleported from nodes A–Z without directly sending quantum states through the channel. This is a crucial application of quantum memories, enabling the creation of large-scale quantum networks for a variety of applications.

3.5.1. *Direct Entanglement Distribution*

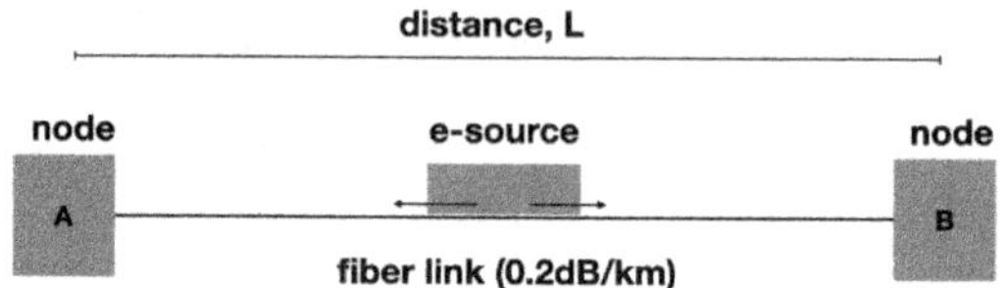

Consider two nodes A and B separated by a distance L, which need to share an entangled set of photons. An entangling source of photons (e-source) such as a spontaneous parametric down-conversion source can probabilistically produce a pair of entangled photons. One photon is sent to A and one to B via an optical fiber channel. The time required to create the entanglement between the two nodes is given by T_{eng} and below we can see the estimation for this entanglement time for three different length scales:

$$T_{\text{eng}} \sim (L/c + T_{\text{s}}) / \eta P$$

$$T_{\text{s}} = 1/\text{rate} \sim 1\,\text{ns}$$

1) $L = 10\,\text{km}$

$$L/c = 30\,\mu\text{s}$$

$$\eta = 10^{-0.02\,L} = 0.63 \qquad \longrightarrow \quad T_{\text{eng}} \sim 4\,\text{ms}$$

$$P = 1\%$$

2) $L = 100\,\text{km},$

$$\eta = 10^{-0.02\,L} = 0.01 \qquad \longrightarrow \quad T_{\text{eng}} \sim 3\,\text{s}$$

3) $L = 1000\,\text{km},$

$$\longrightarrow \quad T_{\text{eng}} \sim \text{eternity!}$$

$$\eta = 10^{-0.02\,L} = 10^{-20}$$

without time multiplexing

Here, c is the speed of light in the fiber, T_s is the timescale at which photons are generated by the source (inverse of photon generation rate), η is the channel transmission and P is the probability of generating photon pairs every time the e-source is triggered. Note that the photon pairs generated in many nonlinear processes such as spontaneous parametric down-conversion or a four-wave mixing process are probabilistically generated. This means that when a source is excited with a pump light, a photon pair may or may not be created. The success probability of such a process is typically a few percent. Although the probability of the generation of one photon pair can increase by increasing the pump power, the probability of generating two, three, and more photon pairs also increases with pump power, which introduces noise to the process. In this case, the state is a mixed state and the fidelity or success rate of faithful quantum communication is reduced.

As L/c is typically much longer than T_s, without multiplexing the communication rate is limited by distance. For different separations, L, the entanglement time is shown above. An optical fiber link has a typical loss of 0.2 dB/km at the telecommunication wavelength. It can be seen that the communication rate exponentially decreases with the distance and for $L > 100 \, \mathrm{km}$, it becomes impractically low.

3.5.2. *Quantum Repeaters Based on QMs*

To overcome the challenges associated with loss, quantum memory can be used to build quantum repeaters. This is done by dividing the channel into multiple smaller segments connected by nodes. By populating each node with sources and memories, the BSM can be carried out between neighboring nodes to share or swap entanglement. It can be shown that in this case, the communication rate decreases polynomially with distance and not exponentially. A simple linear array of memories can overcome the loss, but the probabilistic nature of photon generation remains an obstacle for scalable quantum networks. Parallel processing is additional functionality that can be added to quantum repeaters to further increase the communication

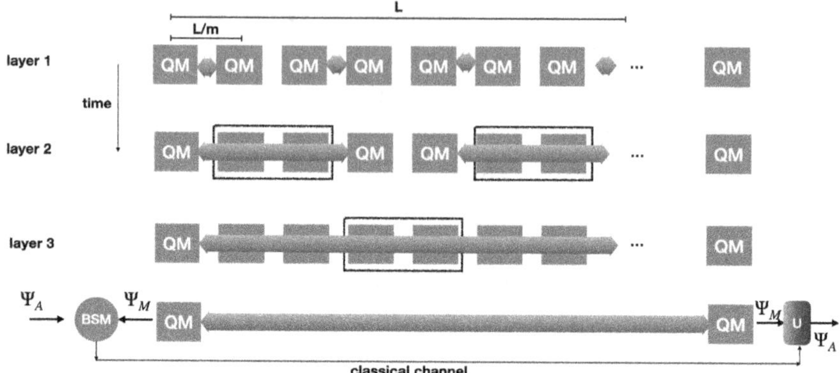

rate and overcome the challenges associated with the probabilistic photon generation process. In parallel or multiplexed processing, each node contains multiple sources (or modes) and memories where entanglement generation and storage can be carried out simultaneously within different channels or different modes.

3.5.3. *Multiplexed Quantum Repeaters*

Multiplexed quantum optical processing refers to the ability to perform BSM on the multiple numbers of modes shared between two nodes. The results are then fed forward to perform subsequent measurements at the next layer, only between the entangled modes established at the first layer. Therefore, multiplexing relies on switching under feedforward control, a crucial functionality that benefits from the coexistence of classical signals in the channel. Consider a small linear system of memories and sources as shown above, where photon sources are generating photon pairs with probability P_0 which are used to create heralded entanglement between two quantum memories. The bipartite entanglement rate can be calculated as $R_e = \frac{c}{L}\eta_t\eta_d^3\eta_m^4 P_0^2$, where η_t, η_d, and η_m are the efficiency of channel transmission, detection, and memory, respectively, and L is the link distance. The transmission efficiency of the fiber link is $\eta_t = e^{(-L/2L_{\mathrm{att}})}$ where $L_{\mathrm{att}} = 22$ km (assuming for 0.2 dB/km fiber loss). The detection efficiency is bounded by the BSM probability, assuming $\eta_d \leq 1/4$. For spontaneous sources, $P_0 \sim 0.05$ is limited

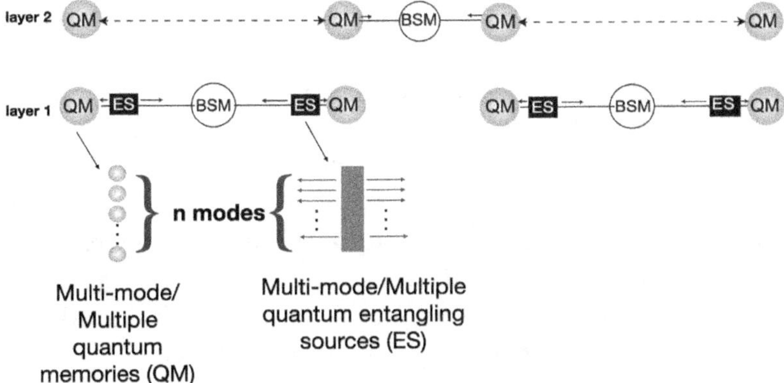

by the probability of having higher-order photon numbers. In a multimode/multilayer network and for k levels of entanglement swapping (nesting layers) in a repeater network, the entanglement rate can be expressed as

$$R_e = \frac{c}{L} \frac{\eta_t \eta_d^{n+3} \eta_m^{n+4} P_0^2}{3^{n+1} \Pi_{k=1}^{n} (2^k - (2^k - 1)\eta_m \eta_d)}.$$

As an example, for four memories ($n = 2$) extending the communication distance to $3L = 1000\,\mathrm{km}$, and assuming memory efficiency of 0.9, one can estimate one entanglement event every 1 hour! Although this entanglement rate is much faster compared with direct fiber links (less than one event every 300 years), in this traditional approach, due to the low entanglement rate and the need for quantum memories with coherence times of an hour, one can conclude that such a network is of no practical use.

To go beyond traditional communication schemes, provided technology is available, one can simultaneously process multiple modes of photons using multimode BSMs, multimode sources, and memories. The photon generation probability can, in principle, approach unity when strong light–atom interactions are employed. Considering the regime of strong cavity interactions, P_0^2 can increase by more than an order of magnitude. Moreover, assuming that M number of spatial, temporal, or spectral modes can be processed in parallel, the communication rate can linearly increase with M. Quantum memory with bandwidth, BW, and storage time T_2 can support up to $\text{BW} \times T_2$ temporal modes.

Instead of parallel processing of multiple modes M, multiplexed processing of entanglement can increase (for certain memory storage times) the communication rate as much as M^{2^k-1} (k is the number of nested layers in the network). Nodes using $M = 100$ modes with only two layers of processing can improve the entanglement rate by a factor of one million. The above figure shows the rate as a function of mode number for various protocols for $L_0 = 100\,\text{km}$. The rate enhancement is more pronounced for longer distances.

Based on the quantitative discussions above, one can identify that some of the most important device metrics in building high throughput scalable networks include the photon generation probability, the number of modes processed, the memory efficiency and the storage time. A plot of heralded entanglement rate (e-rate) for different protocols discussed above is provided below as a function of number of modes.

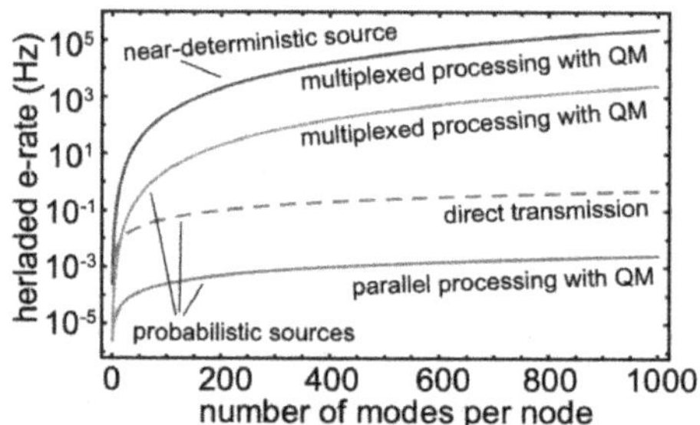

Although it is known that multiplexing and multiplexed processing are essential for future quantum communication, efficient implementation of multiplexing in the network setting is not well understood. The image above illustrates some of the network protocols for communication between two sites, each with multiple nodes and modes. Compared with a switched network where all the resources are exclusively reserved for a specific flow, and linearly multiplexed channels where multiple copies are used to perform parallel processing, the statistical multiplexing protocol is predicted to provide higher throughput. Statistical multiplexing is the quantum analog to a classical internet-based approach, which is more efficient than other approaches. The statistical approach is well suited to dynamic and complex topologies where the service is provided based on the current availability of resources (entanglement). Due to the probabilistic nature of entanglement distribution, the network equally distributes entanglement resources (i.e. photons, BSM, and memories) between multiple routes, and once entanglement is created in a certain route, it signals the user to make use of it. See Aparicio *et al.* (2011) for more details. It was shown that the gap between the performance of a multipath routing protocol and that of a linear chain grows exponentially with distance.

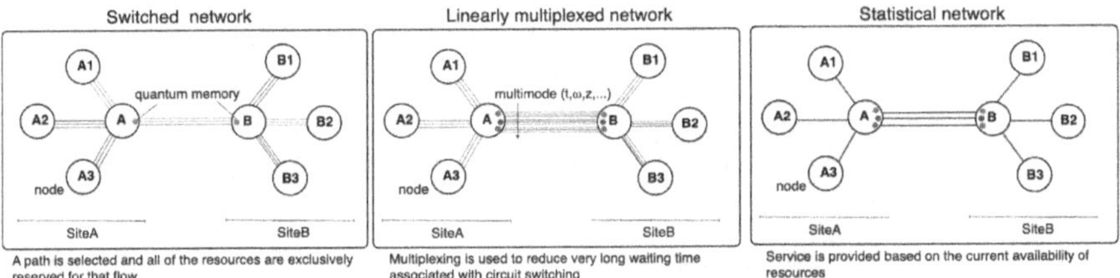

3.6. Quantum Optical Networks: Implementation and Applications

At a high level, a quantum network needs to provide quantum communication services in a scheduled and controlled manner to the quantum nodes connected to the network. The hybrid design of a node enables connections to larger-scale communication networks via satellites or quantum repeaters and using quantum memories.

3.6.1. *Hybrid Quantum Optical Interface*

Here, we describe an example of a possible physical layer of a quantum network. As before, we assume that optical photons are carriers of quantum information. They connect different elements of the quantum network, including a quantum processor (a schematic diagram of an atom-based quantum processor shown on the right) and entangling sources and memories (shown on the top). Different devices in the network have to be compatible with each other and

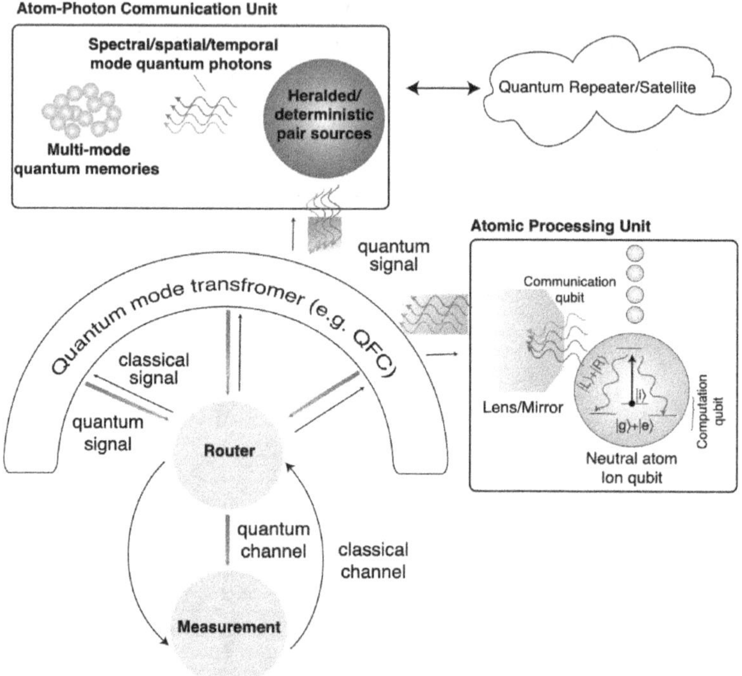

the underlying infrastructure. This requires photons generated and
stored by different devices to have the same temporal, spectral,
spatial, and polarization modes. A practical network will be a
hybrid system consisting of devices with different specs. To ensure
mode matching, mode transformation or mode composition may be
required between photons routing in different parts of the network.
An atom-based quantum processor (e.g based on neutral atoms or
ions) has a direct interface with optical photons. A high numerical
aperture lens placed near an atom or an ion (used for quantum
information processing) can be used to collect photons from the
atomic qubit. The generated photons and the atomic qubit can
be entangled using different protocols. For example, the atomic
qubit can be prepared in the state, $|i\rangle$, and then excited by a
resonant pump to generate a photon with left or right circular
polarization ($|L\rangle$ or $|R\rangle$) upon de-excitation to $|g\rangle$ or $|e\rangle$ atomic
state. The final state of the atom and photon can then be written as

$1/\sqrt{2}(|g\rangle|L\rangle + |e\rangle|R\rangle)$, which is an entangled state. An independent entangling photon source (e.g. a spontaneous parametric down-conversion source) can also generate polarization-entangled photons, albeit with different wavelengths and/or bandwidth. The role of a quantum mode transformer in the network is to ensure different photons' modes are matched before undergoing measurement or storage inside a quantum memory. Quantum frequency conversion (QFC) or bandwidth control are examples of mode transformations.

The role of a quantum router is to decide which photons should undergo joint BSMs and communicate successful BSM results to other layers of the network. A successful Bell measurement will result in the entanglement of nodes with the corresponding measured photons.

Unlike a classical router, a quantum router does not directly route information. In a quantum router, by deciding which channels one should perform BSMs on, the resources are used to create entanglement between nodes at the other ends of the channels. The results of Bell measurements are communicated using classical signals to the corresponding nodes to perform the necessary unitary transformations to complete the entanglement distribution and/or teleportation tasks. Classical signals are also used to control the timing of different processes and resource allocation and to provide necessary feedback controlling drift in the system (e.g. polarization rotation, power fluctuation, etc).

3.6.2. *Routing Information in Multiplexed Networks*

The schematic diagrams in the next page show one example of routing quantum information in a multiplexed optical network. Consider two multimode sources (A and B) of bi-photons. These could be polarization-entangled photon pairs generated at distinct frequency modes, and temporal or spatial modes (mode number: $1, 2, \ldots, m$). A reconfigurable mode switch can control the distribution of entangled photons (see the demonstration of this capability in this Lingaraju *et al.* (2021)) between nodes A and B containing quantum memories optimized for each mode (i.e. $A_1, A_2, \ldots, A_m$ and $B_1, B_2, \ldots, B_m$).

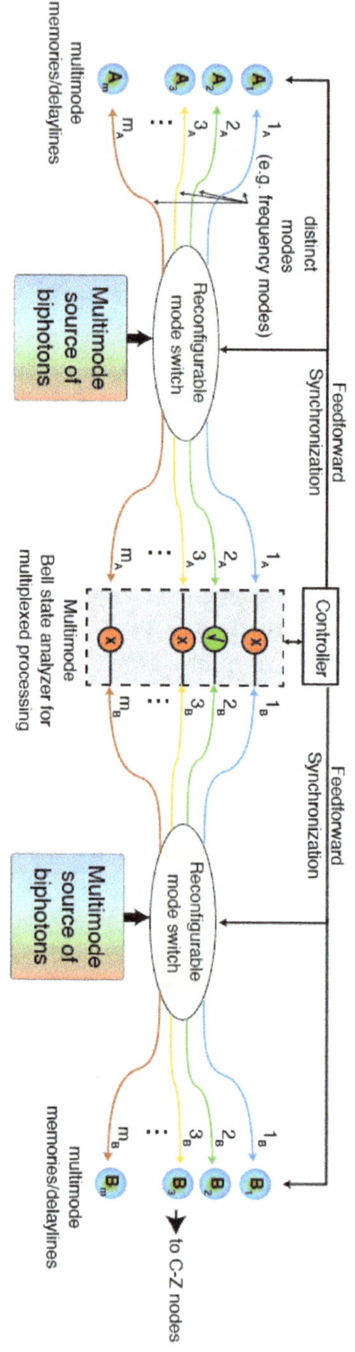

multimode
memories/delaylines
Multimode
source of
biphotons
Reconfigurable
mode switch
distinct
modes
(e.g. frequency modes)
Feedforward
Synchronization
Controller
Multimode
Bell state analyzer for
multiplexed processing
Feedforward
Synchronization
to C-Z nodes

The BSM between photons of the same mode, probabilistically produced by different sources, is performed in the middle node. The result of the Bell measurement is communicated via classical channels to memory nodes and reconfigurable switches used to perform subsequent unitary operations and resource allocation, respectively. This feedforward operation is essential to realize multiplexed information processing. In the example above, joint measurement on mode 2 can result in the successful generation of heralded entanglement between memories A_2 and B_2. At the next layer, Bell measurements are performed between photons stored in memories whose entanglement is established at the previous layer. A similar process can establish entanglement between memories in nodes C and D, and so on. Performing Bell measurements between memories B and C whose entanglement is already determined by the feedforward signal can lead to entangling memories in A and D modes. The process can continue until the end target nodes are entangled.

3.6.3. *Interfacing Spontaneous Sources with Disparate Memories*

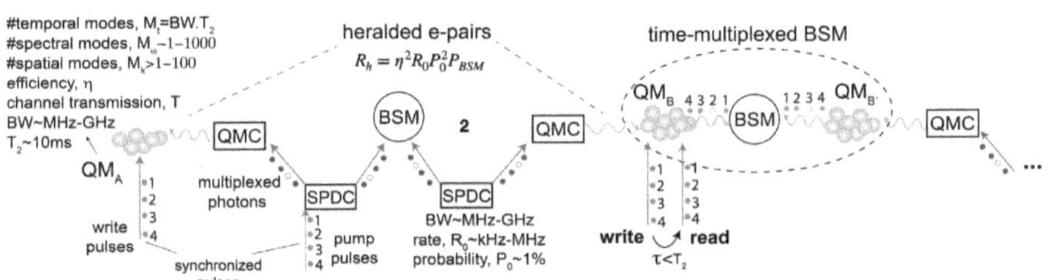

The above image shows a scheme for interfacing photons generated by spontaneous parametric down-conversion (SPDC) sources and atomic quantum memories (QM), highlighting some of the typical performance matrices. Typical SPDC sources and memories are not compatible in wavelength and bandwidth. This is a major bottleneck when interfacing these disparate devices together in a network. SPDS sources typically have a bandwidth (BW) of greater than 100 GHz that can be reduced to MHz using optical resonators to match the bandwidth of an atomic memory (~MHz–GHz). To match the wavelength and/or photonic quantum basis (e.g. polarization basis) of the source and memory, we consider a generic quantum mode convertor (QMC) in the network above between sources and memories. The SPDC and memory can be pulsed synchronously to define distinct and addressable temporal modes (shown as pulses 1,2,3,4). Due to the desire to reduce the higher-order events (and reduce noise) the probability (P_0) of an SPDC source generating a pair of photons given a pump pulse is approximately 1%. The rate at which these photons are created (R_0) depends on the bandwidth and is typically chosen to be about 0.1% of the source bandwidth to reduce temporal mode overlap and thus reduce noise.

A quantum memory has bandwidth limitations that determine the maximum temporal modes (or delay-bandwidth product, M_t) which can be stored. The number of spatial modes stored in the memory depends on the mode capacity of an individual or array of memories. Memory efficiency (η) and coherence time (T_2) can, in principle, reach >80% and >10 ms, respectively and separately. Obtaining both high efficiency and long coherence times can be a challenge, but it is achievable.

One can approximately estimate the maximum heralding rate of entanglement without multiplexing as $R_h = \eta^2 R_0 P_0^2 P_{\mathrm{BSM}}$, where P_{BSM} is the probability of successful BSMs. With multiplexing, P_0 can be replaced with P_M which takes into account the enhancement resulting from all modes.

3.6.4. *Example: Memory-Based QKD*

Here, we review one recent experiment where an atom-like qubit or more precisely vacancy center (VC) in diamond was integrated into a nanophotonic cavity (Lukin *et al.*, 2020) and used as a quantum memory device (panel (a) above). The memory operation and near-deterministic atom–photon and photon–photon entanglement are carried out by sending RF and optical pulses to the atom–cavity systems when a time-bin photonic qubit interacts with the atom (see panel (b) above). The time-bin qubit can be represented as a superposition of an early and late photon as $|e\rangle + e^{i\phi_s}|l\rangle$, where ϕ_s is the relative phase between the two states. The light confinement by

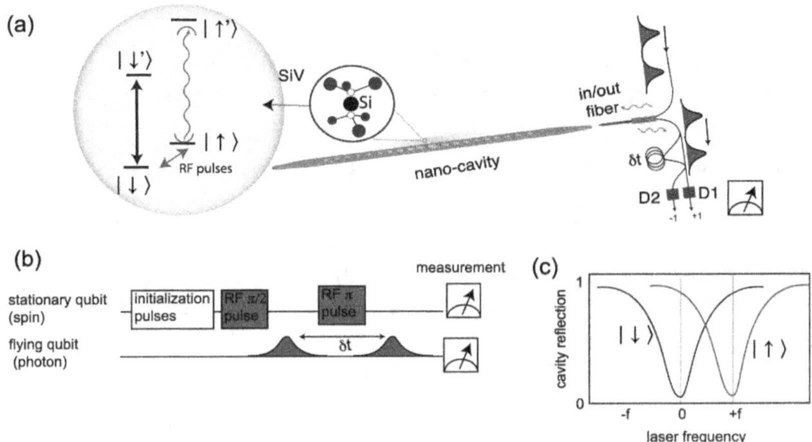

the nanophotonic cavity gives rise to large light–atom cooperativity ($\eta > 100$). Depending on the atomic state of the VC (up or down state, $|\uparrow\rangle$ or $|\downarrow\rangle$), the resonant frequency of the cavity is different as seen in panel (c) above. The larger-than-one single-atom cooperativity induces a cavity shift of about $\delta\omega = \eta\kappa\frac{\Gamma}{4\Delta}$, where Δ is the frequency difference between the photon and the atomic transitions and Γ is the excited state linewidth. Depending on the state of the VC, Δ is different resulting in different cavity resonance shifts. In this scenario, a photon with frequency 0 (referenced to the center frequency of the atom) is only reflected if VC is in the "up" state.

Consider a photonic time-bin qubit: $1/\sqrt{2}(|e\rangle + e^{i\phi_s}|l\rangle$, impinging on the atom–cavity system and let us follow the experimental sequence shown in panel (b) to see how entanglement can be created in this example. First, the VC is initialized by optically pumping it to the $|\downarrow\rangle$ state. Then a $\pi/2$ RF pulse resonant with $|\downarrow\rangle \rightarrow |\uparrow\rangle$ transition is applied to perform unitary transformation and create the following atomic state: $1/\sqrt{2}(|\uparrow\rangle + |\downarrow\rangle))$.

Ignoring normalization for now, the state of VC photon, before interaction, can be written as

$$|\Psi\rangle = (|e\rangle + e^{i\phi_s}|l\rangle)(|\uparrow\rangle + |\downarrow\rangle)$$

$$= |e\rangle|\uparrow\rangle + |e\rangle|\downarrow\rangle + e^{i\phi_s}|l\rangle|\uparrow\rangle + e^{i\phi_s}|l\rangle|\downarrow\rangle$$

Right after the photon in the early time-bin (with center frequency "0"), i.e. $|e\rangle$, interacts with the atom–cavity system, it will only be reflected (remains in the channel) if the atom is in state $|uparrow\rangle$, while the late photon state will be unaffected as it has not interacted with the system yet. The VC photon state then becomes

$$|e\rangle\,|\uparrow\rangle + e^{i\phi_s}|l\rangle\,|\uparrow\rangle + e^{i\phi_s}|l\rangle\,|\downarrow\rangle$$

A subsequent RF π pulse performs the rotation ($|\downarrow\rangle \rightarrow |\uparrow\rangle$ or $|\uparrow\rangle \rightarrow |\downarrow\rangle$) on the atomic state and changes the state to

$$|\Psi\rangle = |e\rangle\,|\downarrow\rangle + e^{i\phi_s}|l\rangle\,|\downarrow\rangle + e^{i\phi_s}|l\rangle\,|\uparrow\rangle$$

Now, let us consider a later time when the late-photon state, $|l\rangle$, interacts with the atom–cavity system. Again, the late photon will only be reflected (with high probability as determined by the high cooperativity) if the atom is in the $|uparrow\rangle$ state. The VC photon state after interaction then becomes

$$|\Psi\rangle = |e\rangle\,|\downarrow\rangle + e^{i\phi_s}|l\rangle\,|\uparrow\rangle$$

After interactions, the photon state is detected in a delay-line interferometer. The path of reflected light from the cavity is split in two, inducing a relative time delay of δt equivalent to the time difference between the early and late photons. By placing two detectors, $D1$ and $D2$ after the interferometer, one can make a distinction between states $|X_\pm\rangle = |e\rangle \pm |l\rangle$ (depending on which detector registers a click).

Detection of a photon after the interferometer in the X basis: $|X_\pm\rangle = |e\rangle \pm |l\rangle$ is equivalent to performing a projective measurement on $|\Psi\rangle$, written as

$$\langle X_\pm|\psi_r\rangle = |\downarrow\rangle \pm e^{i\phi_s}|\uparrow\rangle$$

Following the same procedure, the final state after interactions with two time-bin qubits, separated by time t shorter than the VC decoherence time, becomes

$$|\Psi\rangle = |\downarrow\rangle \pm e^{i(\phi_s+\phi_i)}|\uparrow\rangle$$

where subscripts s and i represent signal and idler photons. Now, instead if we choose $\phi_s + \phi_i = 0,\ 2\pi$ (which is experimentally accessible), the measurement of the atomic spin in the X basis collapses the state of the two photonic qubits into a Bell state, e.g. $|e\rangle_i|e\rangle_s + |l\rangle_i|l\rangle_s$. This means that if after the interaction, one finds the atomic state in state $|\downarrow\rangle\pm|\uparrow\rangle$, it suggests the two time-bin qubits are entangled. Note that the detection of the photonic qubit above is done to quantify entanglement and once the system is calibrated, the detection of the VC spin in the X basis is sufficient to herald entanglement between the two photonic qubits.

The example above shows how near-deterministic entanglement between two photons can be achieved using strong interactions with atoms. The long-lived atomic qubits and large cooperativity are important for scalable entanglement distribution.

3.6.5. *A Quantum Network Example Consisting of Users, Servers, and Routers*

An example of a three-node quantum optical network relying on quantum memories for multiplexed communication and independent quantum e-pair sources is shown above. Users A and B are equipped with multimode entangling sources and quantum memories. A server receives a request for $\Psi\rangle$ and performs a BSM between copies of $\Psi\rangle$ and half of the entangled pairs. Multiple BSMs within the quantum router determine which photon modes of user and server are successfully entangled without a need for switching. Unitary rotations at the user stations upon receiving classical information complete the teleportation of the state to users. In this scheme entanglement, a service is provided to users through statistical multiplexing protocols. When multiple spectral modes are supported by broadband memories and multiple spatial modes available in the chip-scale devices, quasi-deterministic entanglement distribution can be tested.

Integration of quantum sources and memories is challenging as most spontaneous sources (e.g. SPDC) have optimal bandwith and wavelengths which are very different from that of long-lived quantum memories. For long-distance communication, telecom photon entanglement is important and sources with wavelength outside the telecom region can be integrated with quantum frequency conversion

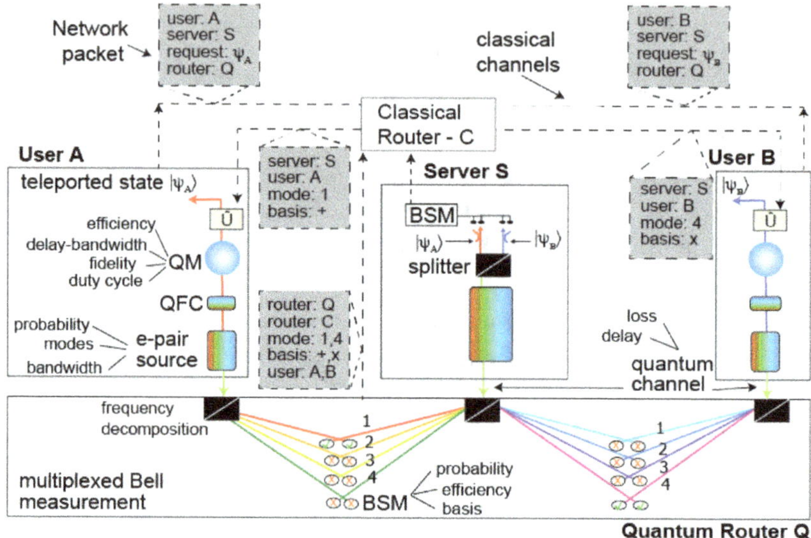

(QFC) device to coherently change the wavelength to/from telecom wavelength. The number of modes and the probability of e-pair generation in quantum sources are other metrics which must be considered when designing a source. A high number of spectral/temporal/spatial modes in which e-pair photons can be generated can increase the communication rate linearly or exponentially with the mode number, depending on whether parallel or multiplexed processing is used. All commercially available sources until 2022 are spontaneous sources with e-pair generation, probability of about 1%. This probabilistic approach to photon entanglement severely limits the communication rates in multi-node networks.

Efficiency, fidelity, bandwidth, center wavelength, storage time, scalable integration, and the duty cycle of quantum memories are important metrics to consider when designing them. Atomic gas memory (in particular, laser-cooled atoms) can provide long storage time, high efficiency, and high fidelity. The bandwidth, non-telecom-band operation, and difficulty with scalable integration are limitations of such memory devices. Solid-state memories such as rare-earth crystals have the potential for long storage time, telecom operation, integration, high bandwidth, high fidelity, and efficiency.

However, the typically high magnetic field and mK temperatures required for operation limit the practical system integration. Also, due to the long preparation times of both atomic gas and solid-state memories, the memories are not continuously available for storage, which again affects the overall communication rate.

Research is underway to address the challenges in both atomic gas and solid-state memories, and the topology and architecture of networks described in the context of the above examples are subject to change or complete transformation as future technologies come online.

Bibliography

Afzelius, M., Simon, C., De Riedmatten, H., and Gisin, N. (2009). Multimode quantum memory based on atomic frequency combs, *Physical Review A* **79**, 5, p. 052329.

Aparicio, L., Van Meter, R., Meyers, R. E., Shih, Y., and Deacon, K. S. (2011). Multiplexing schemes for quantum repeater networks, in *Quantum Communications and Quantum Imaging Ix*, Vol. 8163, doi: 10.1117/12.893272.

Bhaskar, M. K., Riedinger, R., Machielse, B., Levonian, D. S., Nguyen, C. T., Knall, E. N., Park, H., Englund, D., Loncar, M., Sukachev, D. D., and Lukin, M. D. (2020). Experimental demonstration of memory-enhanced quantum communication, *Nature* **580**, 7801, pp. 60–+, doi: 10.1038/s41586-020-2103-5.

Bowen, W. and *et al.* (2003). Experimental investigation of continuous-variable quantum teleportation, *Phys. Rev. A* **67**, p. 032302, doi:10.1103/PhysRevA.67.032302, http://prola.aps.org/abstract/PRA/v67/i3/e032302.

Clausen, C., Usmani, I., Bussieres, F., Sangouard, N., Afzelius, M., de Riedmatten, H., and Gisin, N. (2011). Quantum storage of photonic entanglement in a crystal, *Nature* **469**, 7331, pp. 508–511.

Duan, L. M., Cirac, J. I., and Zoller, P. (2002). Three-dimensional theory for interaction between atomic ensembles and free-space light, *Physical Review A* **66**, p. 023818, doi:10.1103/PhysRevA.66.023818, http://link.aps.org/pdf/10.1103/PhysRevA.66.023818.

Duan, L.-M., Giedke, G., Cirac, J. I., and Zoller, P. (2000). Inseparability criterion for continuous variable systems, *Phys. Rev. Lett.* **84**, 12, p. 2722.

Duan, L.-M., Lukin, M. D., Cirac, J. I., and Zoller, P. (2001). Long-distance quantum communication with atomic ensembles and linear optics, *Nature* **414**, p. 413.

Eisaman, M. D. *et al.* (2005). Electromagnetically induced transparency with tunable single-photon pulses, *Nature* **438**, p. 837.

Fleischhauer, M., Imamoglu, A., and Marangos, J. P. (2005). Electromagnetically induced transparency: Optics in coherent media, *Reviews of Modern Physics* **77**, p. 634.

Fleischhauer, M. and Lukin, M. D. (2000). Dark-state polaritons in electromagnetically induced transparency, *Physical Review Letters* **84**, p. 5094.

Gerry, C. and Knight, P. (2005). *Introductory Quantum Optics* (Cambridge University Press).

Gorshkov, A. V. and et al. (2007). *Phys. Rev. Lett.* **98**, pp. 123601–4.

Grosshans, F. and Grangier, P. (2001). Quantum cloning and teleportation criteria for continuous quantum variables, *Phys. Rev. A* **64**, 1, p. 010301, doi:10.1103/PhysRevA.64.010301, http://pra.aps.org/abstract/PRA/v64/i1/e010301.

Harris, S. E., Field, J. E., and Imamoglu, A. (1990). *Physical Review Letters* **64**, 10, p. 1107.

Heshami, K., England, D. G., Humphreys, P. C., Bustard, P. J., Acosta, V. M., Nunn, J., and Sussman, B. J. (2016). Quantum memories: Emerging applications and recent advances, *Journal of Modern Optics* **63**, pp. 2005–2028, doi:10.1080/09500340.2016.1148212, https://www.ncbi.nlm.nih.gov/pubmed/27695198.

Hradil, Z. (1997). Quantum-state estimation, *Phys. Rev. A* **55**, p. R1561.

Hradil, Z. (1997). Quantum-state estimation, *Physical Review A* **55**, pp. R1561–R1564, doi:10.1103/PhysRevA.55.R1561, https://link.aps.org/doi/10.1103/PhysRevA.55.R1561.

Hradil, Z., Summhammer, J., and Rauch, H. (1999). Quantum tomography as normalization of incompatible observations, *Phys. Lett. A* **261**, pp. 20–24.

Husimi, K. (1940). Some formal properties of the density matrix, *Proc. Phys. Math. Soc. Jpn.* **22**, pp. 264–314.

Jeong, H., Ralph, T. C., and Bowen, W. P. (2007). Quantum and classical fidelities for gaussian states, *J. Opt. Soc. Am. B* **24**, p. 355.

Jing, B., *et al.* (2019). Entanglement of three quantum memories via interference of three single photons. *Nature Photonics* **13**(3): p. 210–213.

Lee, N., Benichi, H., Takeno, Y., Takeda, S., Webb, J., Huntington, E., and Furusawa, A. (2011). Teleportation of nonclassical wave packets of light, *Science* **332**, 6027, pp. 330–333, doi:10.1126/science.1201034, https://science.sciencemag.org/content/332/6027/330.full.pdf; https://science.sciencemag.org/content/332/6027/330.

Lei, Y., Asadi, F.K., Zhong, T., Kuzmich, A., Simon, C., and Hosseini, M. (2023). Quantum Optical Memory for Entanglement Distribution, arXiv:2304.09397 [quant-ph].

Li, L., Dudin, Y. O., and Kuzmich, A. (2013). Entanglement between light and an optical atomic excitation, *Nature* **498**, 7455, pp. 466–469, doi: 10.1038/nature12227.

Liao, S.-K., Cai, W.-Q., Liu, W.-Y., Zhang, L., Li, Y., Ren, J.-G., Yin, J., Shen, Q., Cao, Y., Li, Z.-P., Li, F.-Z., Chen, X.-W., Sun, L.-H., Jia, J.-J., Wu, J.-C., Jiang, X.-J., Wang, J.-F., Huang, Y.-M., Wang, Q., Zhou, Y.-L., Deng, L., Xi, T., Ma, L., Hu, T., Zhang, Q., Chen, Y.-A., Liu, N.-L., Wang, X.-B., Zhu, Z.-C., Lu, C.-Y., Shu, R., Peng, C.-Z., Wang, J.-Y., and Pan, J.-W. (2017). Satellite-to-ground quantum key distribution, *Nature* **549**, 7670, pp. 43–+, doi:10.1038/nature23655.

Lingaraju, N. B., Lu, H.-H., Seshadri, S., Leaird, D. E., Weiner, A. M., and Lukens, J. M. (2021). Adaptive bandwidth management for entanglement distribution in quantum networks, *Optica* **8**, 3, pp. 329–332, doi:10.1364/OPTICA.413657, https://opg.optica.org/optica/abst ract.cfm?URI=optica-8-3-329.

Lukin, D. M., Dory, C., Guidry, M. A., Yang, K. Y., Mishra, S. D., Trivedi, R., Radulaski, M., Sun, S., Vercruysse, D., Ahn, G. H., and Vuckovic, J. (2020). 4h-silicon-carbide-on-insulator for integrated quantum and nonlinear photonics, *Nature Photonics* **14**, 5, pp. 330–+, doi:10.1038/s41566-019-0556-6.

Lukin, M. D. (2003). Colloquium: Trapping and manipulating photon states in atomic ensembles, *Reviews of Modern Physics* **75**, p. 457.

Mikhailov, E. E., Novikova, I., Havey, M. D., and Narducci, F. A. (2009). Magnetic field imaging with atomic rb vapor, *Optics Letters* **34**, 22, p. 3529.

Nurhadi, A. I. and Syambas, N. R. (2018). Quantum key distribution (qkd) protocols: A survey, in *2018 4th International Conference on Wireless and Telematics (ICWT)*, pp. 1–5, doi:10.1109/ICWT.2018.8527822.

Peng, A., Johnsson, M., and Hope, J. J. (2005). Pulse retrieval and soliton formation in a nonstandard scheme for dynamic electromagnetically induced transparency, *Physical Review A* **71**, p. 033817.

Pirandola, S., Andersen, U. L., Banchi, L., Berta, M., Bunandar, D., Colbeck, R., Englund, D., Gehring, T., Lupo, C., Ottaviani, C., Pereira, J. L., Razavi, M., Shaari, J. S., Tomamichel, M., Usenko, V. C., Vallone, G., Villoresi, P., and Wallden, P. (2020). Advances in quantum cryptography, *Advances in Optics and Photonics* **12**, 4, pp. 1012–1236, doi:10.1364/AOP.361502, http://opg.optica.org/aop/abstract.cfm?UR I=aop-12-4-1012.

Poizat, J.-P., Roch, J.-F., and Grangier, P. (1994). Characterization of quantum non-demolition measurements in optics, *Ann. Phys. Fr.* **19**, 3, pp. 265–297.

Qian, L., Lo, H., Youn, S., and Lvovsky, A. (2011). A balanced homodyne detector for high-rate Gaussian-modulated coherent-state quantum key distribution, *New Journal of Physics* **13**, 013003, http://iopscience. iop.org/1367-2630/13/1/013003.

Ralph, T. C. and Lam, P. K. (1998). Teleportation with bright squeezed light, *Phys. Rev. Lett.* **81**, p. 5668.

Saglamyurek, E. *et al.* (2011). Broadband waveguide quantum memory for entangled photons, *Nature* **469**, p. 512.

Saglamyurek, E., Jin, J., Verma, V. B., Shaw, M. D., Marsili, F., Nam, S. W., Oblak, D., and Tittel, W. (2015). Quantum storage of entangled telecom-wavelength photons in an erbium-doped optical fibre, *Nature Photonics* **9**, p. 83.

Sudarshan, E. C. G. (1963). Equivalence of semiclassical and quantum mechanical descriptions of statistical light beams, *Phys. Rev. Lett.* **10**, p. 277.

Tanji, H., Ghosh, S., Simon, J., Bloom, B., and Vuletič, V. (2009). Heralded single-magnon quantum memory for photon polarization states, *Physical Review Letters* **103**, p. 043601.

Tanji-Suzuki, H., Leroux, I. D., Schleier-Smith, M. H., Cetina, M., Griera, A., Simon, J., and Vuletić, V. (2011). Interaction between atomic ensembles and optical resonators: Classical description, *Advances in Atomic, Molecular and Optical Physics* **60**, pp. 201–237.

Thompson, J. K., Simon, J., Loh, H., and Vuletić, V. (2006). A high-brightness source of narrowband, identical-photon pairs, *Science* **313**, p. 74, doi:110.1126/science.1127676, http://science.sciencemag.org/co ntent/313/5783/74.

Tomaello, A., Bonato, C., Da Deppo, V., Naletto, G., and Villoresi, P. (2011). Link budget and background noise for satellite quantum key distribution, *Advances in Space Research* **47**, 5, pp. 802–810, doi:10. 1016/j.asr.2010.11.009.

Trommsdorff, H. P., Corval, A., and Vonlaue, L. (1995). Spectral hole-burning — spontaneous and photoinduced tunneling reactions in low-temperature solids, *Pure and Applied Chemistry* **67**, 1, pp. 191–198.

Wengerowsky, S., Joshi, S. K., Steinlechner, F., Huebel, H., and Ursin, R. (2018). An entanglement-based wavelength-multiplexed quantum communication network, *Nature* **564**, 7735, pp. 225–+, doi:10.1038/ s41586-018-0766-y.

Wigner, E. (1932). On the quantum correction for thermodynamic equilibrium, *Phys. Rev.* **40**, p. 749.

Wootters, W. and Zurek, W. (1982). A single quantum cannot be cloned, *Nature* **299**, pp. 802–803.

Index